CW01073047

PROCUREMENT AND FINANCING OF MOTORWAYS IN EUROPE

RESEARCH IN TRANSPORTATION ECONOMICS

Series Editor: Martin Dresner

RESEARCH IN TRANSPORTATION ECONOMICS VOLUME 15

PROCUREMENT AND FINANCING OF MOTORWAYS IN EUROPE

EDITED BY

GIORGIO RAGAZZI

Università di Bergamo, Italy

WERNER ROTHENGATTER

University of Karlsruhe, Germany

ELSEVIER

JAI

Amsterdam – Boston – Heidelberg – London – New York – Oxford
Paris – San Diego – San Francisco – Singapore – Sydney – Tokyo

ELSEVIER B.V.
Radarweg 29
P.O. Box 211
1000 AE Amsterdam,
The Netherlands

ELSEVIER Inc.
525 B Street, Suite 1900
San Diego
CA 92101-4495
USA

ELSEVIER Ltd
The Boulevard, Langford
Lane, Kidlington
Oxford OX5 1GB
UK

ELSEVIER Ltd
84 Theobalds Road
London
WC1X 8RR
UK

First edition 2005

British Library Cataloguing in Publication Data
A catalogue record is available from the British Library.

ISBN-10: 0-7623-1232-7
ISBN-13: 978-0-7623-1232-0
ISSN: 0739-8859 (Series)

∞ The paper used in this publication meets the requirements of ANSI/NISO Z39.48-1992 (Permanence of Paper).
Printed in The Netherlands.

CONTENTS

LIST OF CONTRIBUTORS

Monika Bak — Department of Comparative Analysis of Transportation Systems, University of Gdansk, Sopot, Poland

Germà Bel — CIPA, Cornell University, Ithaca, New York, USA and Departament de Política Econòmica, Universitat de Barcelona, Barcelona, Spain

Chiara Borgnolo — TRT Trasporti e Territorio, Milano, Italy

Svein Bråthen — Molde University College, Molde, Norway

Jan Burnewicz — Department of Comparative Analysis of Transportation Systems, University of Gdansk, Sopot, Poland

Xavier Fageda — Departament de Política Econòmica, Universitat de Barcelona, Barcelona, Spain

Alain Fayard — French Ministry for Infrastructure and Transport, Paris, France

Carlos Fernandes — Instituto Superior Técnico, Departamento de Engenharia Civil e Arquitectura, Lisboa, Portugal

Francesco Gaeta — French Ministry for Infrastructure and Transport, Paris, France

Andrea Greco — Università di Bergamo, Bergamo, Italy

Peter Mackie — Institute for Transport Studies, University of Leeds, Leeds, UK

Rico Maggi — Università della Svizzera Italiana, Lugano, Switzerland

Marco Ponti	Politecnico di Milano, Dipartimento di architettura e pianificazione, Milano, Italy
Emile Quinet	Emeritus Professor, Ecole Nationale des Ponts et Chaussees and Paris School of Economics, France
Giorgio Ragazzi	Università di Bergamo, Bergamo, Italy
Pippo Ranci	Università Cattolica, Milano, Italy
John Hugh Rees	European Commission, Bruxelles, Belgium
Eoin Reeves	Department of Economics, University of Limerick, Limerick, Ireland
Werner Rothengatter	Institut für Wirtschaftspolitik und Wirtschaftsforschung, University of Karlsruhe, Karlsruhe, Germany
Roman Rudel	Università della Svizzera Italiana, Lugano, Switzerland
Malcolm Sawyer	University of Leeds, Leeds, UK
Nigel Smith	Institute for Transport Studies, University of Leeds, Leeds, UK
Ornella Tarola	Università della Svizzera Italiana, Lugano, Switzerland
Fabio Torta	TRT Trasporti e Territorio, Milano, Italy
José M. Viegas	Instituto Superior Técnico, Departamento de Engenharia Civil e Arquitectura, Lisboa, Portugal

INTRODUCTION

This book is a selection of papers presented to the international conference "Highways: Costs and Regulation in Europe", held in Bergamo on the 26–27 November 2004. The Conference was organised by the University of Bergamo, and sponsored by the European Investment Bank.

We found that there are so many challenging and controversial issues with motorway finance in Europe that it is worthwhile to present them in a comprehensive publication to an international readership. A main outcome of the discussion at the Conference was that a general appraisal bias can be identified in the following sense: Experts from countries which have introduced concession schemes for managing and financing of their motorways are very critical with these schemes, stressing on their shortcomings and caveats. Some of them even follow that a public management under a regime of welfare maximisation would be desirable as a sustainable solution. Experts from countries, which have been sticking to public procurement and tax finance of motorways, strongly attack the inefficiency of public planning regimes and the general tendency to allocate the revenues from special transport-related taxes to the public budget, eventually spending them on other purposes rather than transport. A typical feature of the latter regime seems to be the chronic lack of funds for infrastructure investment.

Looking at the regimes prevailing in Europe it appears that there is a wide variety of possible approaches and of country experience as well. France, Italy, Portugal, Spain and some new member states (NMS: e.g., Hungary,[1] Slovenia) have established concession companies to manage and finance their motorways. Norway has developed some kind of tradition with founding concessions companies for the inter-urban road projects. Furthermore, city road tolling is applied although congestion is comparatively low, for European standards. Austria, Germany and Switzerland have introduced electronic tolling systems for heavy goods vehicles (HGV) on their motorways. Some NMS as, for instance, the Czech Republic and Hungary have introduced access charges by introducing time-based vignettes. In other EU countries such as Sweden, Finland, Denmark, The Netherlands, Belgium and Luxembourg the Eurovignette regime for HGV $\geq$ 12 t is still in place, which sets uniform charges for predefined time periods. In the UK and in some NMS, extended tolling schemes are being prepared which may include light goods vehicles and eventually cars as well apply to a wider road network. Presently it appears that the treasury in the UK will be the responsible institution for finance, while the public administration will operate the roads, assisted by concessionaires for the electronic payment system. This seems to be a remarkable concept for a country, which is regarded most advanced with respect to the privatisation of network economies. Necessary to add that beyond the network-wide public scheme DBFO-finance[2] for single projects is still an option.

It is a most important issue of the book to study country experiences in some detail. Out of 17 papers, 10 are "country studies", describing history, structure, financing and regulation of national highways networks. The regimes exhibited in Table 1 show very different features in the countries where they are implemented. Furthermore, learning effects have developed over time such that the contracts of today look significantly different from those decades ago. From these cross-country and cross-time comparisons powerful conclusions can be derived with respect to the construction of contracts, which allow for enough incentives to private concessionaires and ensure that public goals are not severely violated.

The concession regimes in Spain, Portugal, France and Italy have a different history, which in the case of Italy dates back to the 1930s. Indeed, the primary reason for concession procurement appears to be a financial one: When budget deficits are large and growing, as in recent years, public funds for investments in infrastructures are limited and recourse to private capital seems the answer. The privatisation of Italy's Autostrade in 1999, for instance, was entirely motivated by the wish to raise cash to reduce public debt, as argued by Greco and Ragazzi. If financial considerations become

Table 1. Existing Regimes in Different European Countries.

Regimes	Countries	Papers
Concession companies	France, Italy, Portugal, Spain, Norway	Fayard, Ranci, Greco and Ragazzi, Torta, Ponti, Viegas and Fernandez, Bel, Braaten
Public administration, tolls existing	Austria, Germany, Switzerland, NMS	Rothengatter, Rudel, Bak and Burnewicz
Public administration, Eurovignette system	Scandinavian EU countries, Benelux	Rees, Borgnolo and Rothengatter
Public administration, toll system prepared	UK, Ireland, NMS	Mackie and Sawyer, Reeves, Bak and Burnewicz

paramount, there is the risk of distortions in investment decisions: Authorisations to build new highways financed by private capital (through tolls) are easily granted, while little is invested on ordinary roads to save public funds, even if congestion there may be higher.

There is however no clear tendency towards an increasing role of concessions in Europe; for instance, in Spain, where there are both free and toll highways, the government plans to expand the network of free highways and there is even a discussion about the possibility for the State to "buy back" present concessions, as illustrated by Bel. The coexistence of free and toll highways, in any country, tends to cause regional conflicts, as users in regions subject to tolls feel penalised; tolls are perceived as an additional form of taxation. This is an open question in Spain and to some extent also in France and Italy.

The French "philosophy", that for each highway there should be an alternative state road, so that tolls may be justified by the "better service" offered by the highway, does not seem convincing: To avoid duplications and over-investments highways are planned everywhere as an integral part of the road system, and the value of the "better service" would depend entirely on the quality of the available free alternative.

Concessions also may cause fragmentation in the system. The logic of concessions, whereby each concessionaire should recover over time its investment through tolls and hand back the infrastructure for free to the State at the end of the concession, seems to conflict with a socially optimal pricing policy. If tolls are set to recover total costs of particular network segments the resulting toll structure nationwide may be quite heterogeneous. Optimal pricing would be easier to apply if there was only one concessionaire for the whole country or at least for each macro region, but this would reduce or even eliminate competition among concessionaires, except for the tendering phase (see Ragazzi).

The EU requires that, at the end of any concession period, a new concession be assigned through a public tender. It is not clear which should be the terms of such tender: The new concession could be assigned to the one who offers to run the new concession at the lowest toll, or to the one who offers to pay (to the State) the highest fee for the concession and maximises the tolls under this constraint.

Most motorways concessionaires are highly profitable today, in countries with large traffic volumes and a well-developed highways network, like Italy or France, and also in Spain or Portugal. Instead, countries in the initial stage of highway construction, like Poland or Ireland, find it difficult to raise private capital. The case of Ireland (see Reeves) and the Bebremi case study in Italy (see Greco and Ragazzi) are interesting for the difficulties and length of time needed to negotiate a BOT project. This is to underline that the problem of too long planning phases not only occurs with public regimes of procurement.

Private investors require high returns on equity, even if risks are often covered by the government. For this reason, Ragazzi argues that, if concessions are chosen for budgetary reasons, it would be better to have a small number of state-owned companies to operate the entire system.

The so-called Private Finance Initiative in the UK, where shadow tolls are paid by the government to concessionaires (see Mackie's paper), is criticised by Sawyer mostly for this reason: He argues that, given the risks assumed, private capital has a much higher cost than public funds. Protagonists of private finance will counter-argue, however, that in the case of public finance the risks are hidden and allocated to the taxpayer. Private finance will at least make these risks and their costs transparent.

The scope for competition among concessionaires is extremely limited: Regulation is essential to limit monopolistic extra-profits while insuring efficiency and adequate investments. The quality of regulation is most important for the success of a regime.

Some regulatory issues are common to the whole transport sector, as illustrated by Ponti; regulating tariffs through price cap raises problems similar to the experience of other sectors (see Ranci's paper). Considering the difficulty of writing complete contracts for long periods of time (often concessions are granted for over 30 years) and the risks of "capture" of the Regulator by the licensees, Ragazzi arrives to the radical view that it would be preferable to assign concessions to a government-owned company, or, even better, to "unbundle" concessions, i.e. to have a public agency collect tolls and assign investments and specific services to private companies through tenders.

This idea, which seems radical from the viewpoint of countries with powerful concession companies, has been realised in Austria. In 1982, Austria has founded a state-owned enterprise called ASFINAG to manage, operate and finance the motorways and expressways. In 1997, a vignette system has been introduced to finance these roads from user charges. As ASFINAG operates under private law the management of the company has some entrepreneurial freedom. This has fostered the commercialisation of the Austrian motorway network, in particular after the introduction of electronic tolling for HGV (Rothengatter). In Switzerland, the whole road network is tolled for HGV, not only the motorways (Rudel). Furthermore, about two-thirds of the revenues from HGV tolling are spent on the large tunnel investments of the Gotthard and the Lötschberg, which means that heavy cross subsidisation is a Swiss practice. In Switzerland the custom's office is collecting the tolls and allocates the revenues to the defined destinations. This scheme is not possible in such a stringent way in the EU countries, because of the European law. Some kind of cross-subsidisation is accepted in the revised form of the EU Directive on HGV tolling in the way that mark-ups on the standard tolls are allowed and the budget collected from mark-ups is allocated to rail investments. But these are exceptional cases and not the rule, like in Switzerland.

In Germany, a state-owned enterprise (VIFG) has been established which collects the toll revenues from HGV tolling and allocates the revenues according to fixed shares to the roads (50%), to the railways (38%) and to the inland waterways (12%). VIFG has no planning competence and is not able to apply a financial management, i.e. to take credits or emit shares. For private investors there is a possibility to engage in project finance for single projects; this is an option, which is intensively promoted by the Ministry of Transport. The main reason in Germany to change to a financing system based on tolling is the massive lack of funds although the revenues from transport specific taxes is three times the budget for road investment, maintenance and operation. Because of the predominant un-economic and politically governed rules for allocating funds in the road sector there are many stakeholders to call for a radical change of the institutional regime towards concessions (e.g., The Scientific Advisory Council of the Ministry of Transport). Strong arguments are put forward to substitute the state's inertia by private management (Rothengatter).

Against the background of the contrasting issues – to adjust concession regimes to public goals on the one hand and to insert private management incentives into network-wide publicly governed schemes on the other, – the European Commission faces the big challenge to harmonise the tolling and charging policies in the Union, to minimise the possible distortion of competition on the road freight transport market, stemming from different country-based tolling schemes (Rees). Directive 1999/62 EC is the legal base for the Austrian and German electronic tolling schemes for HGV and after the revision of this Directive, which has been agreed in April 2005 by the EU Council, all forthcoming tolling schemes for HGV or major changes of existing schemes will be subject to this Directive. This holds for any type of regulatory regime, i.e. for future concessions, public schemes or public/private partnerships. The task to find a compromise in the environment of highly diverging interests of the member states was very tedious for the Commission and the Ministries of Transport. The resulting regulation of toll setting for goods vehicles on TransEuropean Networks will be critically commented (Borgnolo and Rothengatter).

Beyond all heterogeneity and all criticism because of its imperfections the new Directive sets a common framework for both, concession and public regimes. This might be a chance for the systems to converge, such that a major disadvantage of concessions, namely the fragmentation of networks as well as the problems with government failures in public regimes can be overcome.

NOTES

1. Hungary was the first NMS to start with a concession regime. After the negative experience with the M1 project this policy has been changed and a vignette system has been introduced.
2. DBFO: Design, build, finance, operate.

Giorgio Ragazzi
Werner Rothengatter
Editors

THE REGULATORY ISSUES OF TRANSPORT INFRASTRUCTURES, AND OF TOLL HIGHWAYS IN PARTICULAR

Marco Ponti

A SUMMARY OF THE BASIC THEORY

The traditional "social choice" approach states that public intervention is needed in the presence of social goals and/or market failures. This intervention has historically assumed the command-and-control form, via direct production, or, more frequently, by means of public agencies. The presence of "capture," "rent seeking," and "informative rents" phenomena[1] implies poor performances of these agencies, leading to and motivating new approaches based on public regulation mechanisms. Command and control, and regulation and market competition, in turn, can be seen within a "subsidiarity"[2] context: the former is to be employed whenever the latter fails to deliver.

The idea of regulation can be suggested as "State intervention, aimed to reach welfare goals, by setting rules for providing incentives to efficiency-oriented actors." According to this definition, the State has implicit difficulties in merging welfare and efficiency objectives. Furthermore,

Procurement and Financing of Motorways in Europe
Research in Transportation Economics, Volume 15, 1–14

ISSN: 0739-8859/doi:10.1016/S0739-8859(05)15001-X

public enterprises may well be efficiency-oriented actors, although this "orientation" is much more sharp and focused in private profit-motivated firms.

The fact that the State faces problems in getting productive efficiency seems inherently quite natural. On the one hand, the minimization of labor costs is an all-important factor of efficiency, while welfare objectives are, in general, oriented toward enhancing employment and labor conditions. On the other hand, managerial skills are compensated and motivated by profit more than by simple "good governance," which is the best possible outcome of public management.

The very first issue is to define the proper scope of State intervention. Within the transport sector, there is wide range of situations where this intervention is needed: natural monopolies, external costs,[3] information asymmetries,[4] and the existence of incomplete markets. Income distribution can also be included in the scope of State intervention: although it cannot be defined as a "market failure," it can be a legitimate public objective.

For transport infrastructures, which are natural monopolies, the main issue is related to the choice between "command-and-control" policies and regulatory interventions. As seen, within a classical "social choice" model, the public "principal" is assumed, in fact, to be both benevolent and all-knowing, and is perfectly able to obtain efficient results from the "agents" (public companies). Further, the objectives remain strictly and unwavering aimed at welfare maximization. But an assumption of public principals as "humans," and not angels, seems much more realistic[5].

Regulation, even if assumed as the dominant strategy, plays a role limited to a well-defined subset of public objectives: productive efficiency (mainly due to the above-mentioned conflict of interests) and allocative efficiency (due to the presence of natural monopolies and other market failures). It can be objected that other public objectives, such as distributive and environmental issues, cannot be kept strictly at a technical level (i.e., measured in terms of social surplus losses or gains), given their mainly political nature; also, in these cases, a regulatory attitude seems more effective than direct state intervention.[6]

Environmental issues are in theory allocative failures (social surplus is not maximized due to excessive consumption). The concept of externality itself implies a relevant distributive content as some actors damage other actors without due compensation, but the incertitudes linked to the measurement of the related economic costs leave a wide space to political judgment. Note also how, in this case, efficiency cannot be neglected, as the social costs involved in every environmental policy have to be minimized. Again, a regulatory approach looks by definition more efficient: "vouchers" and tariff techniques seem far more promising than the "traditional" approach of imposing constraints and prohibitions.

In conclusion, while the space for public decision remains very large within the transport infrastructure sector, the space for "command-and-control" practices (as an alternative to public regulation) seems to be shrinking, at least in theory.

Toll highways are an important example of these regulatory issues: while public planning is not under discussion for their general layout and location, building, operating, and toll structure and goals are subjects of much controversy and debate, both at technical and political levels.

REGULATION OF INFRASTRUCTURES

The Main Tools of Public Regulation

Transport infrastructures are not only natural monopolies, but also legal monopolies, because they are a relevant "building brick" of land use, planned under a command-and-control type of public intervention. Nevertheless, their operations and physical construction can be efficiently regulated through efficiency-oriented actors (basically private ones). It is already so for pure construction activities, regulated by competitive tendering. Construction and operation ("project financing") practices deserve a more in-depth analysis, as will be seen later.

A proper regulatory regime for infrastructures is a highly complex task, with many aspects still to be tested and even fully understood. Further, the "resistance" from political actors[7] to pass from "command-and-control" regimes to regulation practices seems to be strong.[8]

A wide range of regulatory policies is possible; the subsequent paragraphs discuss the main ones, in the order of their degree of innovative content, i.e., in inverse order of their "distance" from status quo. This also can be seen as another kind of "subsidiarity chain".[9]

Privatization of the Assets

This is the radical British "model" for almost every public utility sector. Nevertheless, the implicit risk for public interest seems quite high, given the "option value" embedded in this choice, which is basically nonreversible. "Capture" risks remain paramount, given both the length of the public–private relationship involved (practically eternal) and the power held by a (generally) large private monopolist, so created by a public decision.

In railways, the U.K. experience has shown severe problems both in information control during the privatization phase[10] and in the subsequent regulatory policy. The core issue is that a private natural monopoly is contestable as a property (others may buy it), but keeps too much power against its public regulator (this policy again assumes a "benevolent, all-knowing prince"). Airport regulation seems to be less critical, even if long-run development questions remain unsolved, due to the presence of more complicated problems (e.g., land-use structure).

However, nowhere within the toll road sector has an asset-privatization scheme been proposed; in the U.K., in particular, there is a strong tradition of free roads, stemming from the historical aversion to the tolls imposed in the middle ages by land owners for using their "private" roads.

Competitive Tendering of Concessions ("Demsetz Competition[11]")

For infrastructure operations, this experience is still quite limited, but in theory it seems a "balanced" policy, limiting the risks of "capture" linked with public–private relationships of long duration. For some type of infrastructure, nevertheless, the length of the concession has to be fine-tuned, referring to the technical content of the assets involved, and the consequent need of sufficient "learning" time for the new-entrant company[12]; it is quite obvious that retaining the same operator for a long time raises the risks of information asymmetries and "capture" phenomena. The need for amortization of long-life investments can lead to infrastructural concessions for longer periods. Yet, this is a highly questionable argument for transport infrastructures and for toll highways in particular: these assets (essentially civil works) have a practically infinite life, and therefore there is no physical amortization at play, only financial amortization (if this is the case); thus, sound contractual constraints on maintenance standards and obligations seem a sufficient controlling tool.

Building and Operating Concessions ("Project Financing")

Current practice for new investments sets long concession periods, assuming the need for a complete recovery of the invested capital. This practice merges the responsibilities of construction, operation, and maintenance, with the consequent overall optimization of the entire "system." However, this approach has to be considered with prudence, both for the "capture" risks implicit in longer duration concessions and for its capability of disguising public expenditures for private ones, via too generous risk guarantees in favor of the private investors (in fact transforming these investments into risk-free "sovereign loans"[13]).

Toll highways are in fact a favorite area for "project financing" schemes within the transport sector, and indeed technical advantages may emerge. But these advantages have to by weighted, as we have seen, case by case against the above-mentioned risks: the political and administrative contexts seem to be the main area of investigation, as along with the financial constraints of the administration involved (a severe shortage of funds may provide incentives for a broader use of this tool).

Tariff Regulation

Tariff regulation is relevant both for allocative efficiency, when there are distributive, congestion, or environmental issues, and productive efficiency, when it has to be reached without periodic competitive tendering (i.e., when the provider of the service is assumed unchangeable, and also in the "extreme" case of privatization of main assets). "Price-capping"[14] is the main technical tool in these cases. We will discuss some specific technical issue on this topic later, concerning in particular toll highways.

Yardstick Competition ("Tournament")

This strategy seems to be the most conservative policy among the ones considered here, and remains quite close to command-and-control practices. The regulatory scheme works by comparing the results of different public companies in the same field setting prizes and punishments in accordance to their performances. When many operators are at play this approach is basically coincidental with a command-and-control policy.

The troubles related to insufficient incentives, mixture of efficiency and welfare objectives, "capture," etc., remain in full at present. Regulators and regulated subjects are not sufficiently separated and juxtaposed. It is then necessary to guarantee a high level of autonomy of the different companies from the central regulator in order to minimize the "capture" risks.[15] The "regionalization" process itself can be seen as a form of yardstick competition, where, even within a command-and-control structure, every region becomes both "residual claimant" for the resources involved and may well compare the result of other regions.[16]

Some Technical Examples of Regulatory Issues within the Infrastructure Sector

Congestion Charging and Allocative Efficiency in General

Congestion implies a mismatch of demand and supply of transport infrastructure (access rationing is basically the same issue). The main problem is related to "project financing."

The rationale of the construction costs of natural monopoly being charged to the users can be related with congestion charges; otherwise charge generates a welfare "dead-weight loss." In turn, congestion charges are assumed to be, by definition, efficient, and therefore the related revenues can be efficiently[17] used for financing infrastructure costs. As infrastructures "suffer" from indivisibilities, in general they are underutilized at the beginning of their technical life and congested toward the end. Nevertheless, the financial needs are the contrary, being maximal at the beginning and declining toward the end of the concession.

This is another element that suggests maintaining a prudent attitude toward "project financing" strategies: the "old-fashioned" competitive tendering of construction contracts, followed by a sound periodic tendering of concessions for operations and maintenance, may often be a more prudent choice, where even the charges to the users can be kept under better control.

Let us see in detail the main case where this problem is relevant: tariffs for toll highways.

As we have seen, an "optimal" tariff (i.e., surplus maximizing), by definition, concerns both productive and allocative efficiencies. However, distributive issues should also be taken into account.

At present, only productive efficiency is pursued by tariffs, and this fact is rather puzzling, since the highway system shows large externalities. It is sufficient to consider a new toll route parallel to a congested road: the tariff aimed at recovering the building cost of the new road (i.e., aimed at productive efficiency) will divert part of its traffic to the congested existing one, leaving the new route underutilized.

This issue is a very realistic one: simulation of a case in Northern Italy shows that the "efficient" tariff, in terms of minimization of overall transport costs in a heavily congested area, is actually lower than the existing one on the toll route crossing that area. Obviously, in different circumstances, where the nontoll alternatives are less congested, or where there are modal alternatives, the "optimal" congestion toll is more traditionally quite high, and well above the cost–recovery level.

But at the extreme, if there are no significant alternatives (i.e., the demand is very rigid), even in the presence of heavy congestion the "efficient" toll can be zero (in the sense that if the demand is rigid, congestion is efficient, generating minimal surplus losses).

This is in order to only demonstrate how far a tariff can aim at productive efficiency compared with one that aims at optimizing the traffic flows on a complex, multimodal network.

Other aspects concern: (a) the environmental externalities, very large for the motorized traffic that plies on these infrastructures. Nevertheless, this type of externality, as well as those related with accidents, are better dealt with at their core, i.e., respectively, emissions (via the gasoline taxation) and insurance prices and the level of traffic violation fines (giving correct "signals" to the worse performers), as recommended by the European Commission.[18] (b) Another issue concerns the trade-off between the "deadweight loss" of a cost-recovering tariff (aimed at productive efficiency) and the marginal opportunity cost of public funds,[19] different from country to country, which is impossible to elaborate properly here.[20]

Nevertheless, this issue provides the opportunity to reconsider the whole "standard" marginal cost pricing approach for infrastructure charging set directly against average cost pricing: cost recovery goals have little relation with the latter approach.[21]

Elaboration on distributive issues is also left out of this summary, even if it is far from being irrelevant: why the perceivers of the benefits of a new highway are not supposed to pay for them? This issue, in turn, is intermingled heavily with electoral aspects: politicians tend quite naturally to allocate public funds where the consensus problems result in a dominant one.

The "Minimal Efficient Dimension" Issue

This is a kind of preliminary issue in regulating network infrastructures,[22] where there is no market pressure to determine their efficient dimensions. The efficient dimension has to be minimal in order to avoid capture risks and the consequent excessive power of the regulated against the regulator. So, the problem is to balance the possible economies of scale against the "excessive power of the regulated."[23]

Toll highway networks have probably very limited economies of scale, since they are related mainly to the dimension of the maintenance centers.[24] Therefore, it is reasonable to split up the concessions into subsets of a few hundred kilometers each. The concession system as existing in the present experiences seems highly questionable. It is now generally based on a set of toll links, or on a single link to be built and operated. Yet, the traffic structure within dense areas (i.e., in the European context) is mainly short distance, and the demand for mobility is served by the entire local network, of which the toll links are just a component, and not always the largest one in terms of capacity.

In this scenario, a toll level that is aimed only at cost recovery (investment, maintenance), or aimed at productive efficiency at best, tends to be especially far from optimal in terms of allocative efficiency. Congestion and environmental externalities determine an "optimal" allocation of traffic flows that can be far from the one induced by cost-recovery tolls. Considering also the possible economies of scale of maintenance and minor investments, an area-based concession scheme[25] seems a much more sensible strategy. Further, an area-based concession may well include other critical components such as the management of traffic information for emergencies or ancillary activities like parking facilities and public transport prioritization. Also, schemes for shifting the number of available road lanes from one direction to another in peak periods can become a component of a package of activities that conceives the road system of an area as an integrated service or utility. As told, these packages have obviously to be committed under competitive tendering, and the duration of the concession can be kept limited (especially if major investments are assigned separately, under a "normal" competitive bidding process).

Financial Issues

The established rule of setting a proper rate of return for regulated companies is based, in general, on the calculation of the Weighted Average Cost of Capital (WACC) index. This index is needed to remunerate properly the invested capital, especially, but not only, when investments are financed through the tariffs in an explicit way, and not left within the price-cap mechanism (in fact, it can be considered a "normal" return on capital, taking into account also the risk component).

Also, the correct evaluation of invested capital [or Regulatory Asset Base (RAB)] within a concession regime is a highly controversial issue. In the first place, its magnitude has to be kept to a minimum: productive efficiency requires, for capital not less than for labor, that the resources employed are only the necessary and efficient ones.

For example, a highway concessionaire obviously has an interest in maximizing both the RAB and the estimation of the WACC: this in fact will imply that the share of the "guaranteed" return on its original investment through the tolls it is allowed to collect will be maximized, reducing the weight of the overall risk of its total financial commitment.

A conflict of interests often takes place within the public sector: in selling a concession, or in privatizing an existing one, the State may be willing to maximize its revenues, and doing so may permit or even promote an RAB much larger than the minimum technically needed to operate the infrastructure efficiently; yet, this capital can be really of limited amount if the physical assets are kept public. This is especially true for toll road concessions. Nevertheless, the actual price at which the concession is sold can be much higher that the "book value" of the capital required: its price may well represent the discounted value of future expected profits. In turn, if this "sale value" is in some way included within the RAB instead of the "book value," there is a risk of a spiraling and self-induced increase of the values of the entire concession system, given the fact that a "normal" level of profit (the WACC) on capital is guaranteed via the tariff mechanism.

Also, this second "overvaluation" problem may generate a conflict of interests within the public administration if short-term public revenue maximization prevails on efficiency goals and on the defense of the users from undue rents.

As a consequence, the definition of a proper WACC requires special attention: as we have seen, it is necessary to take into account the specific level of risk of every regulated sector. Within toll highway concessions, for example, if the commercial (i.e., traffic related) risks are taken away from the concessionaire by the public regulator, the WACC has to be lowered.

Further, since the value of the WACC depends also on the relative weight of debt over equity of the invested capital, in order to avoid an "opportunistic" composition of the capital structure of the concessionaires, it is advisable to define a target "leverage level," i.e., a predefined level of debt–equity ratio.[26]

Finally, concessionaires that are floated in the stock market deserve special attention from the regulator, which is bound to be extremely transparent and prudent in all its regulatory activities, especially as far as the parameters of the price-cap formula are concerned.

Also, the inflation index, which enters in the price-cap formula, has to be handled with care: there is a tendency to curb its level referring to the "planned" inflation, and not adjusting it based on the real one. But this is an improper tool for addressing efficiency: inflation is an exogenous factor for the regulated company, and efficiency goals have to be addressed adjusting the X parameter, which holds this role by definition.

Further Price-Cap Problems: Patterns and Levels of Efficient Costs

The price-cap mechanism, although by far the better known tariff-regulation tool available for infrastructure concessions, faces several problems, of which a few are summarized here, focusing on the toll highways concessions.

A first problem is related to which type of risk has to be left to the regulated companies. Due to the exogenous nature of demand variations on road transport infrastructures,[27] it seems reasonable to leave to the highway concessionaires only the industrial risks, and not the commercial ones, that are related with the level of traffic. It is the same rationale that allows the regulated company a full recovery of inflation within the price-cap formula: in fact, if a company faces a risk that is outside its control, it has to behave "on the safe side" and, therefore, tends to calculate as a "standard cost" the worst possible outcome.

A second problem is related to the "efficientization" (X) parameter included in the price-cap formula. Even though efficient costs can be known only in a "learning by doing" process, its definition requires an accurate benchmarking. Given the dominance of monopolistic and inefficient "examples," the data derivation is quite complicated.

Even the "speed" at which efficiency has to be obtained (implicit in the X value) has to be estimated taking into account the specific constraints faced by each sector (labor contracts, etc.). Obviously, the price-cap recalculation starting bases, set generally each 5 years, are the costs[28] incurred at that moment by the concessionaire, and not its revenues.[29] This periodic readjustment of the tariff is known as the "claw-back" procedure.[30] This procedure means that extra profits (or losses) incurred by the concessionaire within a regulation interval (5 years) cannot be made permanent by the regulator, which re-states "normal" conditions at the beginning of every regulatory period.

The Regulation of Investments

Price-caps and competitive tendering could automatically guarantee the efficiency of the investments: only the ones capable of generating net profits will be implemented by the regulated company. Yet, the largest part of the transport investments in infrastructures is not profitable in financial terms, even for highways, and are generally decided by the public actors for a set of social objectives. As far as this decision remains outside the autonomy of the concessionaire, it is perfectly legitimate to finance the investments with a public source of revenue (direct transfer) or through an allowed-by-the-state increase in tariffs. The first case is, in general, dominant for railways (and ports), while the second is in use for highways (and airports). Nevertheless, guaranteed investment funds for a profit-oriented subject generate the well-know Averch–Johnson phenomenon.[31] For example, a large highway concessionaire is induced to "suggest" new investments even if not economically justified, since they will be financed anyway through a toll increase spread over its complete network, and this guarantees larger "normal" profits for it (and is generally more than well accepted by the political decision makers).

All considered, large investments in the transport sector have to be kept basically within a command-and-control frame, especially if the benefits of "project financing" schemes are not fully guaranteed. This may well be the case for toll highways, which present low technical complexity.

The "Number of Tills" Problem

The core of the problem can be summarized as follows: how complex the regulatory action has to be? There are some trade-offs: a fine-tuned regulation may be in theory more efficient than a rough one, but tends to be less transparent and leaves less space to the regulated companies to develop general strategies of optimization.

For toll highway concessions, a double till is already present when investments are decided and financed on top of the regulation of tariffs. If tariff regulation also takes into account congestion and environmental issues, we can speak of a "triple till," i.e., three different "tools" of public intervention.

Another issue in this sector concerns the service areas on the highway network, which are now becoming important retailing and restaurant activities (on top of traditional gas stations). A separate regulation of these activities (or sub-concessions) is generally recommended in order to avoid excessive market power of the highway concessionaire (i.e., "vertical integration"), since no economy of scale appears in operating jointly the highway and these facilities.

CONCLUSIONS

Public regulation of transport infrastructures is a highly complex task, and is basically in its infancy. Command-and-control practices dominate even when they are no longer needed. A proper regulatory process in turn is slowed down by extended "capture" phenomena.

Command-and-control practice and regulation have to be considered within a "subsidiarity" approach, defining a hierarchy of strategies. The traditional assumption known as social choice, of a benevolent and all knowing prince, is no longer acceptable, even if the perfectly egoistic prince embedded in the public choice scenario is also too extreme. A balanced attitude is to stay on the safe side: if you can, do not assume the prince as necessarily benevolent and fully informed.

Within the toll highway sector, several specific issues seem to be dominant. These infrastructures have a technical content lower than other transport sectors (airports, railways, ports) and are "living together" within complex networks with nontoll roads, sometimes with similar characteristics and traffic density, and this is a highly specific condition.

Therefore, a mixture of "common" regulatory issues is present: some of them are similar to other transport infrastructures (and utilities in general) and some have to be understood and addressed by looking in depth into this sector's peculiar aspects.

The main issues are related to the proper role of the concessions to private builders and operators, and the correct apportion of risks between the state and the concessionaires, both within the operation-only scheme and when new investments are the main objective.

Finally, the large dimensions of externalities (larger in overall impact than in other transport sectors) suggest also very specific policies, setting the charging scope on a far more complex basis than in other cases.

Technology can help: the rapid development of toll collection systems, no longer physically connected with the infrastructure, may in a few years offer tools for efficient regulation of the entire road traffic on broader areas, and not limited to isolated links.

But the core issue remains the political role of the regulators: if the public policies are not designed in order to minimize the "capture" risk of rent-seeking concessionaires against the broader interest of the users, it will be difficult that innovative solutions will be designed, and especially so within the toll highway sector, capable of generating very large cash flows without much inherent transparency or widespread information of their final destination.

NOTES

1. See Buchanan (1969).
2. A term of widespread use in European Commission policy papers; here it is used in the broader sense of public strategies set at their optimal level, in relation with the possible role of private operators.
3. Both negative (e.g., environmental and congestion costs) and positive (e.g., Mohring effect).
4. Partially related to safety issues.

5. This not only within the radical context of a public choice setting, where the public "principal" is presented as a maximizing egoistic objective standard "homo oeconomicus." Even within a more relaxed setting, where the ex ante unknown mix of egoistic and altruistic objectives may be varied, a prudent attitude pushes toward some skepticism in assuming a pure "benevolent, all-knowing prince" hypothesis.

6. For example, the decision of a region to provide free public transport (while other services are deemed less socially relevant) is a perfectly acceptable choice (but less so if these services are produced via "command-and-control" practices, and not via competitive tendering).

7. See Ponti (2001).

8. Another proof, if still necessary, of the "capture" mechanisms so well defined within the already mentioned "public choice" approach.

9. Note how this logical "chain" is somewhat different and more complex from the one proposed by Gomez-Ibànez in his recent book on the general subject of infrastructure regulation (2003). In fact, "private contracts," mentioned in that text as one of the main categories of regulation, are not common within the transport sector, while other issues seem far more relevant.

10. See Nuti (1997) and CESIT (1998).

11. See Demsetz (1968).

12. For example, rail and air infrastructures may well need concessions for a longer duration than toll highways (which have mainly simple maintenance and toll collection content).

13. This was the case for several highway investment programs in Italy, but many other projects also have similar contents. This effect is not easy to be immediately identified, given the ever-present possibility of reopening negotiations in the long run, out of a competitive context.

14. For the price-cap theory, see Marzi, Prosperetti, and Putzu (2001).

15. Even for the Japanese railway reform (perhaps the largest example of a form of yardstick competition within the transportation sector), the model has been adjusted in order to guarantee a high level of autonomy (see Japan Railway and Transport Review, 1994a,b).

16. In Germany, with the decentralization of local rail services, the DB national rail company had to face the pressure of different, budget-minded regions, and had to provide more efficient services; this decentralization at the end has set in motion even a real competition mechanism, with the rise of new entrants in both public and private sectors.

17. And equitably, congestion being largely a "club" external cost.

18. See, in particular, the High Level Group white paper "Fair and efficient charging for infrastructures."

19. See Ponti (2003).

20. See A. Bonnafus (2004) *Ranking transport projects*. Paper presented at the WCTR, Istanbul.

21. See Rothengatter et al. (1999).

22. Note how the problem can be considered a problem of "horizontal unbundling," as compared with the "vertical unbundling" issue dominant in nontransport sectors.

23. The excessive power may also have a negative impact on the proper working of a Demsetz concession market.

24. But on this issue empirical evidence is limited.
25. See Newbery (1998).
26. See A. Bergantino and D. Piacentino, *Valore e costo del capitale nella regolazione tariffaria delle infrastrutture di trasporto: teoria ed esempi*. Paper presented at the SIET conference in Palermo, 2003.
27. Demand varies according to the overall economic growth of the country and according to national and regional transport policies.
28. WACC included, as "normal profit."
29. The objective of the mechanism is to make the users pay only for efficient costs allowing for factor providing incentives, which is linked to the possible additional profits gained in each 5-year period, known as regulatory lag, by the concessionaire thanks to its efficiency.
30. Strange as it may seem, this obvious statement in important cases, for example, in Italian highway infrastructures regulation, is not fully accepted, with large and undue additional profits for the concessionaires, that so prove themselves perfectly able to "capture" the regulator (also thanks to the "far from minimal" dimensions of the concessionaires).
31. See Averch and Johnson (1962).

REFERENCES

Averch, H., & Johnson, L. (1962). Behaviour of the firm under regulatory constraint. *American Economic Review*, *52*, 139–150.

Buchanan, J. M. (1969). *Cost and choice: An enquiry in economic theory*. Chicago: Markham.

CESIT (1998). *Liberalizzazione e organizzazione del trasporto ferroviario in Europa*. Rapporto di sintesi, Roma.

Demsetz, H. (1968). Why regulate utilities. *Journal of Law and Economics*, *11*, 55–65.

Japan Railway and Transport Review (1994a). *Domestic transport in Japan present and future*. Monograph.

Japan Railway and Transport Review (1994b). *Restructuring railways*.

Marzi, G., Prosperetti, L., & Putzu, E. (2001). *La regolazione dei servizi infrastrutturali*. Bologna: Il Mulino.

Newbery, D. M. (1998). *Fair and efficient pricing and the finance of the roads*. Cambridge: University of Cambridge.

Nuti, F. (1997). Il caso britannico. In: Nomisma (Ed.), *Liberalizzazione e privatizzazione nelle ferrovie europee* (pp. 13–29). Firenze: Vallecchi.

Ponti, M. (2001). The European transport policy in a "public choice" perspective. In: *Proceedings of the 9th World Conference on Transport Research*, Seoul.

Ponti, M. (2003). Welfare basis of evaluation. In: A. Pearman, P. Mackie & J. Nellthorp (Eds), *Transport projects, programmes and policies. Evaluation needs and capabilities* (pp. 139–150). Ashgate: Aldershot.

Rothengatter, W., et al. (1999). *Fair payment for infrastructure use*. Unpublished paper written for the Federal Ministry of Transport by Construction and Housing by the Advisory Council on Transport, Germany, pp. 12–28.

ANALYSIS OF HIGHWAY CONCESSION IN EUROPE[☆]

Alain Fayard

INTRODUCTION

Methods of financing the construction, maintenance and operation of road infrastructures and the organisation of road administrations are closely linked and currently in a state of major change for a number of reasons:

- Increasingly severe budget constraints.
- A trend towards the creation of autonomous agencies, contractualisation and delegated management.
- Development of public–private partnership aimed at releasing new financing sources and enhancement of performance.
- Increased development of services and new technologies, including direct, personalised user services (vehicle and freight management, guidance, etc.) in particular, alongside traffic management, general information and safety. The link between transport and information is becoming increasingly close.
- A trend towards a new balance between the different levels of the political and administrative machinery.

[☆] The opinions expressed in this document are those of the author and not necessarily those of the organisations with which the author is concerned.

Procurement and Financing of Motorways in Europe
Research in Transportation Economics, Volume 15, 15–28

ISSN: 0739-8859/doi:10.1016/S0739-8859(05)15002-1

Concession systems are in widespread use in the road sector in Europe. This article intends firstly to present a clearer picture of the application of motorway concession contracts in the various European countries, and secondly to identify more accurately the difficulties currently encountered by the European road administrations in the utilisation of the concession option. Within this framework, this article will analyse the degree of diffusion of the PPP in the motorway field and its key factors of success. In conclusion, this approach will lead to better understanding of the problems of the European Union (EU) in defining a common framework in the field of road pricing.

On the other hand, it must be stressed that the classifications used (thus some figures) might be discussed against the background of the quick changes in the links between companies, the changes in the distribution of shares and the setting up of innovative financial and legal vehicles. Moreover, this paper deals mainly with EU 15 countries, due to the availability of data. Nevertheless, it must be mentioned that some New Member States have got good experience and/or have made detailed surveys and conducted adequate training. The motorway network in the New Member States was, in 2001, roughly 2,900 km long (Eurostat, 2004); moreover tolls and vignettes are utilised.

THE PRACTICE OF ROAD INFRASTRUCTURE CONCESSION: DIFFERENCES AND SIMILARITIES BETWEEN EUROPEAN COUNTRIES

The European Legal Framework Which Remains to be Supplemented

It is interesting to note that at the European level the concessions are not evoked by the treaties. As the interpretative Communication of the Commission underlined,[1] in Community legislation, the only existing legally binding provision concerning concessions has been related to the public works concessions until now. Actually, the first directive on the public works contracts of 1971, on the one hand, gives the definition of the concessions of public works (which will be taken again successively by all the directives on the matter) as "a contract of the same type as a (public works) contract except for the fact that the consideration for the works to be carried out consists either solely in the right to exploit the work or in this right together with payment", but on the other hand, excludes the concessions from their field of application.[2] Nevertheless, a Declaration by the Representatives of the Governments of the Member States meeting in the Council, concerning procedures to be followed in the field of public works concessions[3] made at the same time, set up non-legally binding procedures of publicity which were included only in 1989 in directive 71/305/CEE.[4]

Contrary to the public works directive, the services directive does not contain a definition of the concept of concession of services.[5] This gap was filled recently by directive 2004/18/CE of 31 March 2004, which, on the one hand, confirmed the traditional definition of public work concession and, on the other, introduced the definition of service concession being "a contract of the same type as a public service contract except for the fact that the consideration for the provision of services consists either solely in the right to exploit the service or in this right together with payment". The same directive however excluded the service concessions from its field of application.[6]

Nevertheless, as the interpretative communication of the Commission on the concessions in Community legislation pointed out, concessions have to be abided by in the provisions of the treaties.

Presently, the Green Paper on PPPs[7] proposes broad lines of a definition ("forms of cooperation between public authorities and the world of business which aim to ensure the funding, construction, renovation, management or maintenance of an infrastructure or the provision of a service") and strategies/ideas on how to face the challenge for the internal market: to facilitate the development of PPPs under the conditions of effective competition and legal clarity. This proposal brings into focus a number of considerations somewhat different from the concessions in the motorway field.

A Very Widespread Legal Instrument in Europe but Also Very Diversified

The first point to note is that a wealth of experience exists in Europe (Bousquet & Fayard, 2001) in the area of motorway concessions: out of a total of 57,542 km of motorways, 21,998 km are under concession (38%). European experience in the motorway concession domain is in fact recognised world-wide.[8] Nevertheless, by analysing in detail the situation in the different European States, one should note the various practices in the field of concession. Table 1 gives an overview of the different situations in some countries compared to the number of franchised kilometres (Fig. 1).

Another difference among European countries which can be noted is the legal nature of the concession-holder companies. As a matter of fact, it also

Table 1. European Practice of Highway Concession.

	Motorway Network	Motorway Network under Concession	Concessionaire Companies			
			Public (km)*	Private (km)	No. of public*	No. of private*
Germany	12,000	4[a]	0	4[a]	0	1[a]
UK	3,476	580	0	580	0	3
Austria	2,000	2,000	2,000	0	3	0
Belgium	1,729	1.4[b]	1.4[b]	0	1	0
Denmark	973	34[c]	0	34[c]	2[c]	0
Spain	10,500	2,610	112.6	2,497.4	1	28
Finland	603	69	0	69	0	1
France	10,383	7,840	6,940	900	10[d]	4
Greece	916.5	916.5	916.5	0	1	0
Italy	6,840	5,593.3[e]	1,201.60	4,391.7	7	17
Luxembourg	130	0	0	0	0	0
Norway	629	550[f]	550	0	26	0
Netherlands	2,300	4[g]	0	4[g]	0	2[g]
Portugal	2,271	1,771	0	1,771	0	11[h]
Sweden	1,450	16	0	16	0	1
Switzerland	1,341.9	8.85[i]	8.85[i]	0	1	0
Total	57,542.40	21,998.05	11,730.95	10,267.10	52	68

Note: Kilometres in operation, 1 February 2004.
Source: PIARC, road administrations website and alia for 2003.
*'Public' means "company controlled by the State and/or local collectivities".
[a]Rostock tunnel.
[b]Liefkenshoek tunnel.
[c]Including 18 km of Great Belt Link Sealand and Funen and 16 km of Oresund Link between Denmark and Sweden.
[d]Figures include two international tunnel companies (ATMB and STRF).
[e]Including 30.2 km of tunnels under concession.
[f]The term 'concession' is used here in its broad sense, insofar as the Norwegian companies have an exclusively revenue collection function.
[g]Including 2 km of Noord tunnel and 2 km of Wijkertunnel (shadow tolls).
[h]Lusoponte (operating two 24-km-long bridges).
[i]Grand Saint Bernard tunnel.

appears that out of a total of 21,998 km of motorways under concession, 11,730 km are managed by the public sector (53%) and 10,267 km by private companies (47%). There are currently 52 state-owned and 68 private concession companies in Europe. The predominant position still occupied by government-owned companies in the motorway concession domain in Europe is an aspect which should be borne in mind.

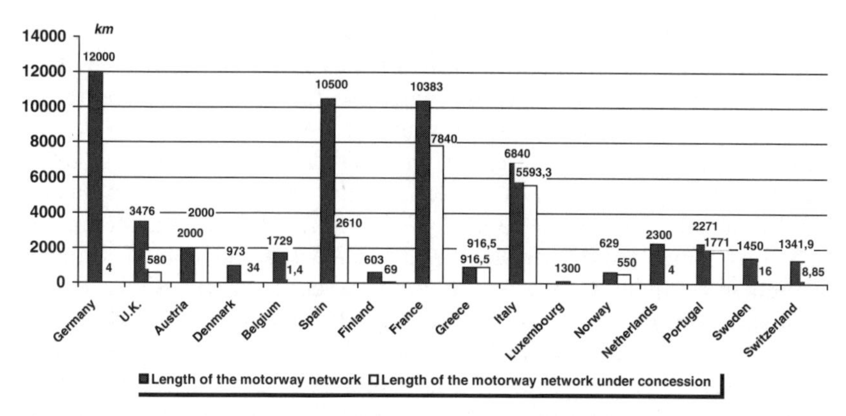

Fig. 1. Overview of European Practices in Highway Concessions (With or Without Toll).

The diversity of the concession systems introduced by the various European road administrations also deals with the respective roles of the concession company and the public authorities. In particular, in the risk sharing between concession authority and concession company, we can note various existing situations.

This question of risk sharing indeed represents one of the major difficulties for road administrations when setting up concession projects. It should be noted here that the increasingly dense motorway network mesh also generates difficulties in the area of commercial risk attribution. The increasing degree of the interrelationship between motorway sections under concession within the same network makes it more and more difficult for the concession companies to carry the commercial risk alone in view of the fact that traffic levels can vary considerably according to commercial policies defined on an individual basis. Consequently, the public authorities will be required to play a regulatory role to a progressively greater extent.

For instance, in France, enterprise contracts, concluded for a duration of 5 years between the State and the concession-holder companies, formalise arrangements for each part as regards work and investments, tariff policy, financial objectives, management indicators, social policy and employment, user service, architectural quality of the works and insertion in the environment. In Italy, Autostrade S.p.a, the largest operator of toll motorways has moved its monitoring system from maintenance planning to total quality management. Performance is defined and periodically measured taking into account the different perspectives of the users (motorists, environment), operator and grantor of the concession as well as the linkage between revenue and performance. Performance measurements are converted into a quality factor that increases or decreases the toll revenue kept by the operator in line with the performance achieved.

This regulatory role is necessary in a context of privatisation: on the Iberian peninsula, there are large construction groups, associated financial establishments, which took control of the national motorway companies: ACS thus controls the Spanish companies Abertis and Ferrovial and the Portuguese company Cintra. In Italy, on the contrary, the State, to avoid possible distortions of competition as regards attribution of work, prohibited the construction groups from owning the principal motorway companies of the country. Thus, family industrial groups were essential: Autostrade, the European leader, is controlled by Benetton, while the Gavio family acquired the Turin–Milan motorways. In France, SASF has been floated in Euronext since 2002; APRR has been listed since November 2004 and SANEF since March 2005.

These companies have started a diversification of their activities (telecommunications, parking lots, etc.) and try to win over new markets (as promoters, concessionaires or services providers). On the other hand, the market is more open and improved efficiency may be expected; nevertheless, it is obvious that competition is mainly for the market and is weak in the market.

There are also differences with respect to concession company selection criteria. European Commission legislation only calls for the obligation of prior announcement for the award of motorway concessions. Once this obligation is met in accordance with prescribed procedures, bids submitted can be freely negotiated. The criteria most frequently quoted by road administrations are the amount of public subsidy required, the credibility of the financial arrangement, the technical quality of the project, operating strategy and price policy, and the reputation of the concession company (inclusion of a construction company amongst its shareholders, etc.). In this respect, the European Commission underlined in its Green Book on the PPP, that "the attribution of particularly complex markets 'requires' to modernize and simplify the Community legislative framework"; it also recognizes that legal insecurity is greater with the public–private partnership when "the absence of clear rules and co-ordinates could in addition be likely to increase the costs related to the installation of such operations" (items 13, 26, 32, 34 of the Green Book on the PPP).

Formulas for determining toll charges also differ throughout Europe ('price cap' method in Italy, traffic band method or availability payment in the U.K., etc.). Each of these formulas corresponds to a particular level of risk sharing, and is consequently of genuine interest for all concession authorities in this regard.

While the functions of toll systems are both numerous and diverse – demand management, regulation, funding, internalisation of external effects, etc. – it appears that road administrations are increasingly confronted with the problem of the social acceptability of road tolls. This depends on five main factors, namely the amount of toll with respect to the type of user, the collection method, the enhancement of user service, the presence of free alternative routes and the possible existence of taxes already allocated to the road sector.

The Directive 1999/62/EC of the European Parliament and of the Council of 17 June 1999 on the charging of heavy goods vehicles for the use of certain infrastructures (to which amendments were proposed by the Commission in a communication COM/2003/0448 final)[9] is mainly oriented towards the internal market aspect. Amendments proposed by the Commission are the reverse of the orientations of the White Paper European Transport Policy for 2010: Time to Decide[10] presented on 12 September 2001, i.e. modal shift from road, innovative approach through PPP and pooling, global charging including external costs. As far as concessions are concerned, the Commission proposed a bureaucratic system of fixing toll and tried to be, to a certain extent, a third party in the negotiation of the concession contract; such a system would prevent the setting up of PPP (because of legal uncertainty) and the development of concession companies (as incentives would be abolished and financial means would be limited). Concession contracts are granted through a competition process and to add administrative approval to market rules results would be counter-productive (without prejudice of enforcement of competition regulation itself). It must be mentioned that the European Parliament's position in its first lecture and the common position adopted by the Council of European Union on 6 September 2005 is more positive.

Last, but not least, concession and toll are two different concepts:

- The concession may be paid either by the user or by the government through a shadow toll or another way of payment such as an availability payment (it is not established that such an acceptance of the term 'concession' is in line with EU legislation but this legislation is of an "obscure clarté").
- A toll may be levied by a public administration such as the Swiss customs (LSVA/RPLP/TTPCP, heavy vehicles fee) or by a service provider such as

EUROPPASS in Austria (Go-Maut for a road operator ASFINAG) or Toll-collect in Germany (LKW-Maut for a public administration, Bundesamt für Güterverkehr).

A FEW SIMPLE IDEAS

Concession and PPP are buzz words but behind these words the contents vary a lot. Labels used in PPP jargon such as turnkey contracts, BOT, DBFO or performance-based maintenance contract have no single and clear definition. Each PPP solution is unique and too complex to be characterised in one word or acronym. There is a continuum of alternatives from works and services contracts to BOT/concessions including performance-based and turnkey contracts.

Except in a few totally state-controlled economies, private firms are always involved in road design, construction, maintenance and operation. But the partnership takes on its real sense when a private firm provides a global service with sufficient autonomy and incentives to produce efficiency gains for the benefit of all parties and in particular road users.

Each PPP has to be designed and the objectives of the policy makers must be taken into account. Each parameter must be adjusted as a sound engineer plays with a synthesiser, the levers of which are:

- scope of work: tasks assigned to the private sector;
- autonomy: initiative left to the private actors;
- pooling: number and type of projects concerned by the agreement;
- risks: how to share them among actors;
- cost recovery: how to pay back, mainly users/tax payers;
- finance: project/corporate finance, government involvement;
- World Bank (2002): http://rru.worldbank.org/toolkits/partnershipshighways.

The sharing of risks between the concession authority and the concession company is a core point of PPP. A concession is of interest to the public authorities insofar as the concession company assumes global responsibility not only for the investment but also for its subsequent management, provided a genuine transfer of risks to the concession company occurs. The fact that operating expenses are just as substantial as construction costs is frequently overlooked. On average, operating costs reach about 75% of construction costs after 35 years.

Furthermore, it is important to draw attention to the problem of the assumption of the commercial risk (toll charge × traffic) in a concession context. In contrast to a simple work contract, the concession company selected by the government bears the financial and technical risks of the investment and carries the greater part of the commercial risk. Nevertheless, this commercial risk is too great in certain instances to be carried by the concession company alone. This is the case in particular where the project is integrated in a meshed motorway network. In this situation, any change in price policy for any part of the network or any addition of a stretch, even if remote from the project under concession, can have major consequences on the traffic levels recorded on the latter. The level of uncertainty concerning traffic predictions for new toll infrastructures is generally high with respect to the duration needed to achieve expected significant dividends, the more so as the estimates cover a lengthy period (concession periods are customarily of the order of 30 years and more). It is therefore advisable to alleviate the commercial risk. The counterpart of such an alleviation may be to set up mechanisms incorporated in the contract between the concession authority and concession company, in order either to cap the amount of toll revenue collected by the concession company, or to control the rate of return of the concession company or to apply a variable concession period. Broadly, the idea is to give more freedom in the fixing of tariffs to the concessionary companies when the network is a grid. On this subject, the practice adopted for DBFO projects, where the commercial risk is controlled by applying a traffic band concept, is very interesting. Alleviation of the commercial risk does not, however, lead to the elimination of any/every risk.

A distinction needs to be drawn between commercial risks, i.e. risks relating to the number of future users, and income levels, with risk-sharing such that all random elements, regardless of the time (design, construction, operation, etc.) at which they may become apparent, will have an adverse impact on the profits and losses account. The final profit and losses account is therefore only partly an outcome of the commercial risks (a certain confusion seems to be in the Commission interpretative communication and in some national legislation as well).

It is stressed that PPP may mean either public–private partnership or public–public partnership. There are a number of advantages to using an autonomous entity rather than a government body to supply services, whether that entity is public, quasi-public or private. The first argument in favour of such an approach is that it makes it possible to increase the funding available for investment by calling on private capital markets without increasing the government debt (under certain conditions). A more convincing argument is that an autonomous entity would be able to design, build and operate an infrastructure more efficiently because it can work on a time-scale that is longer than an annual budget and can operate more flexibly by taking account of the overall cost and by optimising long life cycle investment, maintenance and management of operations.

This autonomous entity does not have to be private in order to be effective. The most important points are stringent accounting practices, a high public image and know-how. It is this same concern to identify responsibilities and offer incentives that provides the basis for policies regarding contractual arrangements between administrative departments that certain countries have adopted.

Creating autonomous entities and introducing contracts allows benchmarking either through peer competition between public and quasi-public entities or, in the case of private enterprises, through market forces. If a private entity is granted exclusive rights and a long-term contract (which is commonly the case with regard to infrastructure), the competition is not in the market but is denied entry to that market and it is important to put in place contracts offering incentives and to limit the rent that the private enterprise can derive from the informational asymmetry from which it benefits. It is therefore extremely useful, if not essential, for the regulator to have references available, i.e. more than one autonomous operator. The use of public–public partnerships (and not public–private partnerships) can offer an effective solution to providing the regulator with reference information (Fayard, 1999).

In its White Paper of 2001, the Commission underlined the important and major implications of the different means of financing infrastructures, in particular through the public–private partnership. The increasingly frequent recourse to private funding for the execution of motorway concession projects must not lead to a withdrawal of governments from the management of road systems. The role of concession authorities is essentially to safeguard the interests of the general public while introducing incentive mechanisms for the concession companies. Experience has demonstrated the importance of the role of the concession authorities in the successful implementation of a concession project, whether upstream (project identification, socio-economic studies to measure the interest of the project for the community, provision of a clear-cut and stable legal framework) or downstream for drafting specifications, negotiating with the candidate concession company and monitoring the concession up to its termination. It is also important to remember that it is only the socio-economic return of a project which provides a relevant indicator for the advantage of an investment for the community.

Table 2. A matrix approach of PPP.

	No Toll	With Toll
Non-shared risks	Public authorities or agency acting directly as employer	Public authorities or agency (may be called 'concession company') acting directly as employer
Shared risks	Construction and/or operation concession Lease or shadow toll (traffic risk?)	Concession (construction and/or operation)
A third dimension: global view of investment maintenance, operation and services		

Finally, it should not be forgotten that, in addition to its task of safeguarding the interests of the community, the concession authority (government) must also concern itself with increasing the awareness of citizens, whether they are users or not. Both for the implementation of the 'user-pay' principle and for the implementation of a concession contract for a project with the private sector, it is of primordial importance, in order to ensure the social acceptability of their decisions, that the authorities take great care to inform the public beforehand of the reasons for their choice. This has the additional advantage of establishing a transparent environment, while associating the public with government decisions to a significant degree and fitting the way of levying a road user charge to social behaviour.

Do not mix up the notions of toll system, concession and private financing (see Table 2):

- In a toll system, the user is charged and not the tax payer. Irrespective of the method of financing used, there are ultimately only two financing sources, these being the actual user and the tax payer; therefore, a shadow toll is not at all a toll and as no additional income is generated; some conceding authorities have to face difficulties for paying the concessionaires. It is possible to mix a toll system and a shadow-toll-type system as in the German A Betreiber Modell; under the A model, the multilane extension of existing motorway sections as well as their maintenance, operation and financing may be assigned to private investors whose remuneration may be backed by LKW-Maut (the German F Modell is a classic financing toll system).
- The introduction of an agency, an autonomous public or semi-public or private entity, frequently in the context of a concession or franchise

arrangement, has the primary advantage of making it possible to impose a management discipline, extending beyond the frequently inappropriate framework of annual budgets, and to institute an organisational system for the maintenance and operation of the infrastructure, and allocate the necessary funding sources. The creation of an autonomous entity is frequently associated with a toll system, but not necessarily (shadow tolls, lease, etc.).
- Private financing assumes risk-taking by the private sector. Nevertheless private financing does not exclude any public contribution, nor is this method necessarily synonymous with a toll system.

Another confusion needs to be avoided: fees or taxes could also be perceived proportionally to distances travelled. That does not involve a concessive structure; in such a case the relevant criterion is to identify the final beneficiary of toll funds: either the public treasury or the concessionaire.

The recourse to tolling can be assessed on the basis of the following three factors:

- the funding constraint which restricts the possibilities for achieving economically profitable investments;
- the allocation of resources, collected from the user rather than the taxpayer, leading to a preference for investments which can be funded to the detriment of other solutions which are more advantageous in terms of the economic results for the community, but which ensure their feasibility[11]; and
- the toll dissuasion effect, which reduces the economic advantage of the development programme.

A toll system can serve to optimise the utilisation of the transport network (traffic spread, inter-modal sharing of traffic load, etc.). In this case, however, charge systems must meet a number of different and sometimes contradictory objectives (marginal cost charging, cost recovery, maximised profit, etc.). Furthermore, an effective pricing policy implies a certain flexibility in toll determination in order to take into account both general goals and specific situations at different levels. This is why until now toll harmonisation undertaken by the European Commission has not borne fruit. The proposal of a common methodology in fixing tolls, a rigid system, has increased the Commission's difficulties.

The increasingly frequent use of private funding must be taken into account when defining the training required by personnel responsible for monitoring concessions. The financial and legal aspects have now taken on a degree of importance such that they must form a genuine part of the basic knowledge of concession authority personnel. Moreover, the bureaucracy involved in funding and infrastructure management has to be reduced by:

- specifying the precise responsibilities of different levels of government and the different operators (public or private) involved;
- encouraging users to acquire a sense of responsibility and to 'own' infrastructure in their own minds, firstly by listening to what they have to say and then by ensuring that users are properly represented (beyond the level of the traditional lobbies); and
- improving the efficiency of operators by reviewing their performance, fostering peer review, bench-marking and developing quality assurance programmes.

This represents a challenge for any public administration.

These actions must not be seen as a privatisation (which they do not exclude in fact), and the word 'commercialisation', which is often used should, not be misinterpreted: commercialisation is one of the possible management tools deployed and is not necessarily the objective. Neither should the public–private partnership be seen as a subtle form of privatisation by osmosis nor should the public–public partnership be seen as bringing down the public administration, but both should be regarded as a genuine alliance in which each side has its own distinct and well-defined role. Clearly, there are a number of political, if not ideological, choices that need to be made, not to say that a cultural revolution must take place.

And last, but not least, what is a 'concession' company: a way of managing facilities in a comprehensive manner and in the long term and/or earmarking resources in a public–public partnership, a commercial company, even listed, but with a systematic distribution of dividends, a way for contractors to have contracts awarded, a company with an actual business plan and the requisite means. In the case of tolls, is the users' charge a recovery of costs or the price of providing a service? Should highways be managed in a regulated market economy or, at the end of the day, is it a natural monopoly de facto under administrative control?

NOTES

1. Interpretative communication of the Commission on the concessions in Community legislation (2000/C 121/02), 29 April 2000.

2. Article 3, paragraph 1 of directive 71/305/CEE of the Council, 26 July 1971, providing for coordination of the implementation procedures of works contracts (OJ L 185 of 16 August 1971).

3. Declaration by the Representatives of the Governments of the Member States meeting in the Council, concerning procedures to be followed in the field of public works concessions of July 26 1971 (*Official Journal* C 082, 16 August 1971, p. 0013–0014, English special edition: Series II Volume IX P. 0055).

4. Council Directive 89/440/EEC of 18 July 1989 amending Directive 71/305/EEC concerning coordination of procedures for the award of public works contracts, *Official Journal* L 210, 21 July 1989, p. 0001–0021, Council Directive 93/37/EEC of 14 June 1993 concerning the coordination of procedures for the award of public works contracts (*Official Journal of the European Communities* L 199, 9 August 1993) and Directive 2004/18/EC of the European Parliament and of the Council of 31 March 2004 on the coordination of procedures for the award of public works contracts, public supply contracts and public service contracts, *Official Journal* L 134, 30 April 2004, p. 0114–0240.

5. Council Directive 92/50/EEC of 18 June 1992 relating to the coordination of procedures for the award of public service contracts, *Official Journal* L 209, 24 July 1992, p. 0001–002.

6. Article 17 of the above-mentioned directive.

7. Green Paper on public–private partnerships and Community law on public contracts and concessions COM/2004/0327 30 April 2004.

8. The European situation differs from the American one where there are few toll motorways ('toll road' or 'turnpike') which, in addition, are mainly built and operated by public authorities. Moreover, the percentage of road traffic for freight is very much lower (28% in the U.S.A., 44% in EU 15).

9. Proposal for a Directive of the European Parliament and of the Council amending Directive 1999|62 EC on the charging of heavy goods vehicles for the use of certain infrastructures (COM (2003)) 448 final.

10. White Paper on European Transport Policy for 2010: Time to Decide, COM(2001) 370 final. 12 September 2001.

11. And one should bear in mind that the socio-economic return varies according to the date of completion.

REFERENCES

Bousquet, F., & Fayard, A. (2001). *Road infrastructure concession practice in Europe*. Policy Research Working Paper, WPS 2675. Washington, DC: World Bank Institute.

Eurostat (2004). *Eurostat*. Luxemburg.

Fayard, A. (1999). *Overview of the scope and limitations of public–private partnerships*. Paris: ECMT.

World Bank. (2002). *A toolkit for public–private partnership in highways*. Washington, DC: World Bank.

CONCESSIONS VERSUS NETWORK-WIDE TOLLING SCHEMES, THE COMMUNITY FRAMEWORK FOR MOTORWAY TOLLING IN EUROPE

Chiara Borgnolo and Werner Rothengatter

INTRODUCTION

This chapter elaborates on two basic options to manage and finance inter-urban roads, in particular the motorways, in Europe: either concession companies are established, which build, operate and finance new parts of the motorway system, or the network is operated and administrated as a whole, presumably by a state-owned enterprise. Financing in the latter case may be realised partly by levying tolls, partly by other fiscal instruments if not all vehicle categories are priced and the revenues from tolling are not sufficient to recover the full costs.

We first present the concession regimes in some countries, mainly in southern Europe. The main ideas of promoting concession regimes are professional management and stable finance. But there are also some caveats to be discussed, as, for instance, the risk of heterogeneous solutions for different network parts, that might be detrimental for spatial competition.

Procurement and Financing of Motorways in Europe
Research in Transportation Economics, Volume 15, 29–40

ISSN: 0739-8859/doi:10.1016/S0739-8859(05)15001-X

In concession regimes, all vehicle categories are included in the pricing scheme and have to contribute to finance the total costs and a return on investment for the concessionaire.

Some countries are favouring network-wide solutions, as, for instance, Austria, Germany and Switzerland (non-EU). In this case, not all vehicle categories are included in the tolling scheme or the tolling schemes might be different for the vehicle categories. For example, heavy goods vehicles may be priced on the basis of kilometres driven while cars may be priced by vignette systems or by fuel taxes.

Since 1999, motorway tolling was legally based on Directive 1999/62 EC. In effect, this Directive sets out the legal framework for new tolling schemes for heavy goods vehicles in Austria and Germany while it did not apply to concession regimes. In the revised version of this Directive, which was agreed on by the Council of Ministers in April 2005, an attempt has been made to design a general framework for tolling heavy goods vehicles on the motorways and eventually the primary road network. Naturally, the new legacy established represents a compromise between various country preferences. Therefore, it is not consistent in every respect, very restrictive with some issues and rather general with others.

MOTORWAY CONCESSIONS

General Characteristics

Distance-based tolls are levied on some 20,000 km of motorway networks mainly located in Southern Europe. Countries which historically relied on toll collection to fund motorway development include France, Italy, Portugal and Spain. More recently, Greece, Croatia and Slovenia have also chosen to levy tolls to fund the development of their national motorway networks. In closed, tolled motorway networks each vehicle is identified (either visually or electronically) as it passes through a toll plaza and then charged (either manually or electronically) as a function of vehicle attributes and distance between entry and exit plazas.

On tolled networks as a whole, the average proportion of revenue generated by car and trucks is 80 and 20%, respectively. This indicates that the tolls are not only based on infrastructure costs but also on the willingness to pay, because the share of commercial vehicles of the total costs of the road infrastructure comes out in most studies between 45 and 55%. In some combination, with the diffusion of pre-paid cards and credit cards, a variety of electronic fee collection (EFC) systems has been introduced in the last decade to charge vehicles without requiring them to stop. All EFC systems in use for toll collection are based on dedicated short-range communication (DSRC) systems with quite simple on board units and roadside equipment located at toll plazas. With an agreement on European standards reached only in the summer of 2001, the first generation of EFC systems has been developed under specification of concerned motorway concessionaires. In countries where different motorways links are operated by several concessionaires, inter-company remote payment procedures have also been progressively developed at national levels – TELEPASS in Italy, T.I.S. in France and Via-VERDE in Portugal – to enable motorway users to use the same payment means on sections operated by different concessionaires and to allocate revenue among the latter as a function of mileage performed in each section.

A Summary of Pros and Cons Concerning Concession Regimes

Concession regimes have already a long tradition such that their strengths and weaknesses are well known. Naturally, the way in which they fulfil private and public expectations depends on the contract between the state and the concessionaire. In particular, the less satisfying experiences are widely following from government failures in the context of contract design or governance. Nevertheless, there are fundamental arguments against concession schemes when it comes to the efficiency of the network management as a whole, i.e. beyond the financial efficiency of the single units. The pros are as follows:

1. The concession company will be interested in efficient management, low construction cost and good service to the customer. It may overcome *X*-inefficiency of public institutions.
2. The issue of financing motorways has been successfully met in most cases. Earmarked revenue from tolling gives a stable source of finance, which is not disturbed by political interference.
3. Acceptance of prices is higher compared with state management; there is higher degree of freedom with setting prices according to willingness to pay and capacity use.
4. Risk (excluding political risk and force majeure) can be included explicitly in the tolls, and risk management can be an issue of the concession company.

The cons are as follows:

1. Capital cost might be higher if the concession company is rated lower than the state in the capital market.
2. As conditions vary with every project the tolls might come out different and provide a heterogeneous pattern over space, which is detrimental to fair spatial competition.
3. Concession companies are spatial monopolies and the relationship with the state is subject to captivity. This may result in suboptimal contracts from the welfare point of view.
4. Concession companies focus in the first instance on financial success. Therefore, their pricing regimes, if derived from entrepreneurial decision-making, might be in conflict with welfare maximising toll setting.

To avoid caveats of heterogeneous spatial pricing patterns public regulation comes in, in some cases in the form of obligatory tariffs and in others in the form of price caps. Obligatory tariffs can be set uniformly in space to avoid heterogeneity. But in this case a major parameter of private management is extracted from the concession regime, namely the management-based setting of prices.

NETWORK-WIDE TOLLING REGIMES

From Time-based to Distance-based Tolls[1]

From the mid-1990s, in almost all European countries where motorway development has been traditionally funded trough general taxation/public budgets, the purchase of a permit was made mandatory for both domestic and international commercial vehicles using national motorways for a given time period (year, month, week or day).

The adoption of vignettes in Europe was pioneered by two EFTA countries – Switzerland and Austria – both determined to cope with quite sustained transit traffic across Alpine motorway road corridors. In Switzerland, the measure was complemented with bans for vehicles over 28 tonnes. In Austria, the adoption of time-based vignette was jointly implemented with ECOPOINT, a scheme conceived to allocate transit quotas against emission targets.

Since 1995, the purchase of the vignette was progressively made mandatory for goods vehicles ‘having a maximum permissible gross laden weight of no less than 12 tonnes' using motorways in Germany, Holland, Belgium, Luxembourg and Denmark. Sweden had joined the Eurovignette club in 1998. Each national scheme is framed in accordance with key provisions of Council Directive 1999/62/EC on the charging of heavy goods vehicles for the use of certain infrastructures.

In Switzerland, since January 2001, a system is in operation to charge commercial vehicles ($\geq$3.5 tonnes) as a function of vehicle attributes and distance performed on the whole Swiss road network. As the whole road network is priced and the tolls include elements of external costs, the tolling regime is not compatible with the EU legislation.

In Austria, since January 2004, an electronic system entered into operation to charge heavy goods vehicles ($\geq$3.5 tonnes) for the use of the Austrian motorway network. Within a ten-year contract, EUROPPASS has been implemented and is operated by a consortium led by Autostrade Spa, the main motorway concessionaire in Italy, which developed the TELEPASS system for EFC at the beginning of the 1990s.

In Germany, since January 2005, the Eurovignette system has been replaced by an electronic charging system for heavy goods vehicles ($\geq$12 tonnes). The payment system is based on GPS and GSM and is not dependent on physical toll plazas. These somehow interrelated tolling systems for heavy goods vehicles are compared in Table 1.[2]

Also in the UK a network-wide tolling system for lorries and other vehicle categories is under preparation.[3] The driver for introducing a British lorry road user charging (LURC) system is not only generating an additional source of revenue. It is anticipated that the charge will be – at least partly – offset by a reduction of fuel tax, which presently is the highest in Europe. The UK government plans to introduce an electronic tolling system; however, the details are not yet decided. Table 2 lists some details of the plans for LURC.

COMMUNITY FRAMEWORK FOR ROAD CHARGES

Directive 1999/62/EC

In the absence of a community framework for motorway concessions, in regions of the European Union both with and without a tradition of tolled motorway concessions, rules on infrastructure charges are limited to those set in Directive 1999/62/EC (see European Union, 1999 and 2005) on the charging of heavy goods vehicles for the use of certain infrastructures. The

Table 1. Comparison of Network-wide Toll Collection Systems.

	Germany	Austria	Switzerland
Toll requirements	12 tonnes permissible vehicle weight	= 3.5 tonnes maximum permissible vehicle weight	= 3.5 tonnes maximum permissible vehicle weight
Legal basis	Motorway Toll Law for Heavy Commercial Vehicles (ABMG) from 12 April 2002; Ordinance by the Federal Ministry for Transport, Building and Housing (BMVBW) Regulation; Federal Road Toll Law 2002 (BStMG), BGBl I No. 109/2002 from 16 July 2002	Ordinance by the Federal Ministry for Transport, Innovation and Technology (BMVIT) based on the BStMG ASFINAG Toll Ordinance from 1 September 2003	Heavy Transport Tax Law (SVAG) from 19 December 1997 Ordinance by the Swiss Department of Environment, Transport, Energy and Communications (UVEK)
Toll operator	Toll Collect GmbH; www.toll-collect.de	EUROPPASS LKW-Mautsystem GmbH; www.go-maut.at	Regional Customs Office (OZD), Bern; www.zoll.admin.ch
Supervisory authority	Federal Office for Goods Transport (BAG), Cologne; www.bag.bund.de	Motorway and Expressway Financing Corporation (ASFINAG), Vienna; www.asfinag.at	Regional Customs Office (OZD), Bern; www.zoll.admin.ch
Tolled roads	Federal motorways	Motorways, expressways	Complete road network; Toll routes in kilometres
Tolled routes	12,000 km	2,000 km	71,000 km (2.1% motorways, 25.9% main roads, 72% other roads)
Distance-based toll (in km)	From start of toll system, projected: €0.09–0.14 (no turnover tax)	From 1 January 2004: €0.13–0.273 (excl. 20% turnover tax)	From 1 January 2001: €0.11–0.45 (no turnover tax)
Calculation basis	Distance travelled, number of axles, pollution class	Distance travelled, number of axles	Distance travelled, permissible vehicle weight, pollution class
Projected toll income per year	€2.8 billion	€600 million	€509.4 million (CHF 800 million)

Table 1. (*Continued*)

	Germany	Austria	Switzerland
Payment method	Bar, EC/credit card, fuel card, direct debit	Bar, EC/credit card, fuel card	Bar, EC/credit card, fuel card, direct debit
Technology	GPS, wireless mobile (GSM)	DSRC module (microwave, infrared)	Microwave technology, speedometer, GPS, DSRC module (microwave)
Automatic log on	On-board units as needed (projected long-term: up to approx. 800,000)	Approx. 400,000 on-board units	55,000 on-board units
	Internet log-on	400 toll portals	200 toll portals System: enforcement
Manual log-on	3,500 toll station terminals	No alternative log-on available	Payment booths at border crossings (for foreign vehicles)
Enforcement	Approx. 300 control bridges; 278 enforcement vehicles	Approx. 100 control units (integrated into the toll portals) plus 34 mobile teams	Approx. 10–15 control bridges (five currently in operation)
On-board unit Installation	On-board unit (OBU) Domestic trucks: no installation required, manual log-on option available	Go-Box Domestic trucks: installation required/stickers required for trucks = 3.5 tonnes	Tripon CH-OBU 1 Domestic trucks: installation required for all trucks = 3.5 tonnes
	Foreign trucks: no installation required, manual log-on option available	Foreign trucks: installation required/stickers required for trucks = 3.5 tonnes	Foreign trucks: no installation required, manual log-on option available at border crossing stations
Cost	On-board unit: free Installation: costs paid by vehicle	On-board unit: €5 Installation: free (sticker applied by vehicle owner/operator)	On-board unit: free Installation: costs paid by vehicle owner
Distribution/installation	Distribution and installation: *In country:* approx. 1,600 service partners	Distribution: approx. 220 GO sales points (along the roadway, at all key border crossings)	Distribution: Regional Customs Office (OZD), Bern
	Foreign: approx. 350 service partners	Installation: sticker applied by vehicle owner/operator	Installation: approx. 370 authorised garages

Table 2. Indicative Timetables in Adoption of LURC.

Introduction	Procurement
Spring 2004 – Initial legislation	Spring 2004 – Issue a prior information notice
Spring 2005 – Legislation on structure, collection and administration of charges and key definition	Spring 2004 – Publish official journal of EU. Advertisement and pre-qualification questionnaire
2006 – Design and build phase, e.g. installation of roadside equipment	Summer 2004 – Potential supplier open day
2006 – Secondary legislation including regulation	Summer 2004 – Preliminary invitation to negotiate
2006 – Recruitment and training of staff	Contracts awarded by end 2005
2006–07 – Pilots and testing	
End 2006 – Go live for pre-registration services	
From 2007 to 08 – Equipment installed in vehicles and start of revenue collection and fuel duty repayment	

so-called Eurovignette Directive was adopted to complement the creation of a single market for road haulage with a framework to harmonise fixed taxes and infrastructure fees which member states levy on good vehicles 'having a maximum permissible gross laden weight of no less than 12 tonnes'.

The Directive applies to vehicle taxes, tolls and time-based access charges (user charges). While ruling that tolls and user charges may not discriminate, directly or indirectly, on the ground of the nationality of the hauler or origin destination of the vehicle, the Directive:

1. Establishes minimum rates (Euros per year) for fixed taxation components levied by member states – having basically the nature of vehicles excise duties and/or motor vehicle licences.
2. Gives the same definition of the type of network (art 2) where tolls are levied on all motorised vehicles and user charges can be levied on good vehicles that are registered in a different country than the one where the trip is undertaken: 'motorway or dual carriageway road specially designed and built for motor traffic, which does not serve propriety bordering nor it does cross at grade with any road, railway or tramway track, or footpath'. In addition, member states (including those where user charges are levied) can levy tolls for the use of bridges tunnels and mountain passes.

3. Establishes maximum rates for user charges (Euros per year, month, week or day).
4. Defines criteria for setting tolls, i.e. payment of a specific amount for a vehicle travelling the distance between two points of the infrastructure referred in point 2).

Directive 1999/62 EC Revised

The revision of Directive 1999/62 EC has been prepared during a tedious process of negotiations over about 2 years and has been accepted by the Council (not yet by the Parliament) in the spring of 2005. The main motivation was to create a unified platform for motorway tolling in the EU and to update the old Directive in a number of areas to charge heavy goods vehicles in a way free of discrimination, allowing for allocative incentives to improve capacity use and environmental performance in the transport sector. The basic principles are:

1. Tolls shall be based on the principle of the recovery of infrastructure costs. Specifically, the weighted average toll shall be related to the construction costs and the costs of operating, maintaining and developing the infrastructure network concerned. The weighted average tolls may also include a return on capital or profit margin based on market conditions.
2. Member states may vary the toll rates for purposes such as combating environmental damage, tackling congestion, minimising infrastructure damage, optimising the use of the infrastructure concerned or promoting road safety. Toll rates may be varied according to Euro emission class, provided that no toll is more than 100% above the toll charged for equivalent vehicles meeting the strictest standards. Furthermore, toll rates may be varied according to the time of day, type of day or season, provided that no toll is more than 100% above the toll charged during the cheapest period of the day, type of day or season.
3. Toll rates may in exceptional cases for specific projects of high European interest be subject to other forms of variations. For infrastructure in mountainous regions a mark-up may be added to the tolls of specific road sections, which are subject of acute congestions affecting the free movement of vehicles, or the use of which by vehicles is the cause of significant environmental damage. The mark-ups may not exceed 25%.
4. The weight limit, which formerly was set to 12 tonnes, may be reduced to 3.5 tonnes. The tolled network, which previously only included

motorways and roads of similar quality, may be extended to other parts of the road network, which are affected by the tolling scheme.

5. Strict rules are defined for the calculation of infrastructure costs and their allocation to the vehicle categories. The benchmark of pricing is the weighted average costs of the infrastructure. Costs include capital costs (depreciation and interest on capital), structural repair and current costs of operation and maintenance. Calculation of capital costs is only allowed for new infrastructure (age less than 30 years), exceptions from this rule have to be demonstrated by the member state.
6. The calculation of tolls shall be based on actual or forecast heavy goods vehicles' shares of vehicle kilometres adjusted. If they are based on forecast traffic levels a correction mechanism shall be provided to correct any under- or over-recovery of costs due to forecasting errors.
7. Tolls and user charges may not both be imposed at the same time on any given category of vehicle for the use of a single road section. However, member states may also impose tolls on networks where user charges are levied for the use of bridges, tunnels and mountain passes.
8. Discounts may be given to frequent users, not exceeding 12% of the standard toll. Unjustified disadvantages to non-regular users should be avoided. If on-board units are necessary they should be available under reasonable administrative and economic arrangements.
9. Tolls and user charges shall be applied and collected and their payment monitored in such a way as to cause as little hindrance as possible to the free flow of traffic.
10. Member states are free to apply tolls and/or user charges on roads not included in the trans-European road network.
11. For new concession companies established after the transposition of the Directive, the maximum level of tolls shall be equivalent to the level that would have resulted from the use of a methodology based on the core calculation principles of the Directive. Tolling arrangements already in place shall not be subject to the obligations set out in the Directive for as long as these arrangements remain in force and provided that they are not substantially modified.

The main problem behind the tedious process of revision was the different positions of the member states with respect to:

- *Magnitude of tolling*. While the peripheral countries were interested in low tolls in centrally located countries with high traffic volume, the countries in the geographical heart of the EU, exposed to a high volume of transit traffic, were interested in high tolls to control the growth of international traffic demand, in particular of transit traffic.

- *Use of the revenues.* While the Commission and a majority of countries favoured a strict hypothecation of revenues some countries were strictly against (e.g. the UK). In the version approved the use of the revenues is subject to the subsidiarity principle, i.e. up to the decision of the member states.
- *Treatment of externalities.* The Commission and several countries were in favour of integrating external costs in the cost definition. After a long time of negotiations this has been reduced to uncovered accident costs, and finally also this element was cancelled.
- *Special rules of cost calculation.* Some of the special rules of cost calculation are not based on scientific foundations but on political compromise. This might be a weak point for the future process of unification of cost calculations.
- *Applicability to concession regimes.* The member states, that apply concession regimes, were interested in maintaining the arrangements. This holds, in particular, for the tolling structure which in most cases included willingness-to-pay elements that are not based on infrastructure costs. The Commission, however, was interested in a unified legacy for motorway tolling.

CONCLUSIONS

Unfortunately, the Commission was not able to set up a legal framework for road concession companies. This would have been in essence the major challenge to standardise the schemes of operation and finance, to monitor the monopoly power of concession regimes and to define clear rules for the allocation of public/private responsibilities.

In the absence of a legacy for establishing concession regimes for motorways and other roads the revision of the Directive for motorway tolling of HGV shows severe gaps of regulation. While the new Directive 1999/62 EC is binding for public and new concession regimes, while the old concession structures are left unchanged. This results in a rather heterogeneous structure of motorway tolls.

The treatment of vehicles below 12 tonnes is subject to the member states. This means that incentives are driving the road haulers to apply suboptimal technology to save tolls, i.e. to use vehicles with 11.9 tonnes. If the different treatment of old concession regimes and public finance/new concession regimes will be implemented strictly according to the new rules then a – in the social sense-suboptimal routing of trucks will pay for the haulage companies.

To conclude, much effort has to be invested by the Commission to transpose and monitor the Directive according to economic principles as it has not been possible to integrate these principles in the legal framework.

NOTES

1. For the theoretical foundation see Newbery (1998); for the political process see Kageson (2000).
2. See BMVBW 2001 and 2004; Swiss Federal Office for Spatial Development, 2002; ASFINAG internet service.
3. HM Custom and Excise, Department of Transport (2004).

REFERENCES

BMVBW. (2001). *Federal Ministry of Transportation, Construction and Housing. LKW-Maut auf Bundesautobahnen beschlossen*. Berlin, 15 August; Press release number 208/01.

BMVBW. (2004). *Federal Ministry of Transportation, Construction and Housing. Fakten zur LKW-Maut*. Accessed 5 July 2004.

European Union. (1999 and 2005). *Directive 1999/62 EC: Directive of the European Parliament and of the Council on the charging of heavy good vehicles for the use of certain infrastructures OJ L 187, 20.7.1999*. Revision adopted in April 2005.

HM Custom and Excise, Department of Transport. (2004). *Modernising the taxation of the road haulage industry – progress report 3*. London: HM Treasury.

Kagenson, P. (2000). *Bringing the Eurovignette into the electronic age: the need to change Directive 99/62/EC to allow for kilometre charging for heavy good vehicles*. Brussels: T&E European Federation for Transport and Environment.

Newbery, D. M. (1998). *Fair and efficient pricing and the finance of the roads*. Cambridge: University of Cambridge.

Swiss Federal Office for Spatial Development. (2002). *Fair and efficient – the distance-related heavy vehicle fee in Switzerland*. Berne, June.

TOLLS AND PROJECT FINANCING: A CRITICAL VIEW

Giorgio Ragazzi

INTRODUCTION

Until very recently, collecting tolls was a viable possibility only on highways, for the simple reason that entrances are few. The evolution of satellite and computer technology is now opening up the possibility of collecting tolls on any kind of roads and to differentiate among types of vehicles, hour of travel, etc. A 'dream world' is opening up for economists and traffic engineers: it is now becoming feasible to apply marginal social cost (MSC) pricing![1]

Yet, in some European countries, highways are free, whereas in others some or most motorways are subject to tolls. In Latin countries (Italy, France, and to some extent Spain), tolls depend upon the distance and 'class' of vehicle (no difference is made for the time of travel), and their level is set, basically, so as to insure that revenues cover costs (including operator's profit). Due to budgetary pressures, there is a growing tendency to finance new motorways under project financing, thus introducing tolls to cover, as far as possible, all costs. This approach is predicated on the basis of equity (it is 'fair' that users pay for the infrastructure they use) and/or greater efficiency by private concessionaires vis-à-vis public road departments. It is evident, however, that setting the level of tolls to cover costs ('average cost pricing') contrasts with MSC pricing, and may cause suboptimal use of the road network. In this chapter, I will analyse the

Procurement and Financing of Motorways in Europe
Research in Transportation Economics, Volume 15, 41–53

ISSN: 0739-8859/doi:10.1016/S0739-8859(05)15004-5

internal contradictions of the project financing approach to road construction (section 'Average Cost Pricing') and the arguments in favour of greater efficiency by private concessionaires (section 'Efficiency').

In the section titled 'Private or Public?' I review the peculiar difficulties and risks of regulation in this sector, to conclude that it is preferable for concessionaires to remain state owned. In the section titled 'Road Traffic Taxation', I compare taxes to tolls, as instruments to finance roads. Lastly, in the section titled 'Taxing for Congestion', I consider the merits of applying tolls only to control congestion, or to charge for the differential damage to roads caused by different classes of vehicles.

AVERAGE COST PRICING

In the standard project financing approach to the construction of a new highway, the public authority, having defined a project and possibly determined the amount of public subsidy, if any, assigns the concession through a tender to the operator who offers to build, finance and operate the highway, for a given number of years, at the lowest toll. This amounts to setting tolls at a level that covers costs (operating and financial). An example, in Italy, is the Brebemi project (see Torta's chapter in this book).

Governments tend to favour this system mainly because it reduces the need for public funding. The main argument used by politicians to foster the acceptance of tolls on highways (and wherever project financing is used) is that it is 'fair' that users pay for the 'better service'. Various objections may be raised on this point.

The 'better service' is supposed to be compared to 'normal' roads, which are free. But this is obviously a vicious circle: the worse state roads are in, the greater the demand for highways. In Italy, for example, state roads are still basically the old Roman roads (Aurelia, Emilia, etc.). Narrow, no emergency lanes, winding up and down mountains, crossing villages and cities with traffic lights and low speed limits. The enormous growth of road traffic from the 1960s onwards, which yielded an enormous growth in fuel tax revenues, could not have taken place without the construction of highways. Highways are certainly not an optional infrastructure for people who love fast driving. They are planned as an integral part of the road network. Their construction is a substitute for new investments (and thus reduces congestion) in 'normal' roads. If users of highways are called to pay for the full financial cost, they are actually subsidizing users of free roads and general taxpayers. The benefit principle for justifying tolls is clearly misplaced.

Applying tolls only on motorways, at levels set to cover costs, diverts traffic towards 'normal' free roads, whenever an alternative is available, and thus causes suboptimal use of the road network. It also results inevitably in a casual and irrational set of toll levels.

Indeed, a coherent application of the principle that 'users must pay for the cost' would call for cutting tolls to cover only operating and maintenance costs, once financial amortization has been completed. If this principle was applied, the result would be an erratic and irrational structure of toll pricing for different spans of highways, depending upon historic costs and seniority, the more so the grater is the fragmentation of the network among several operators.

If a new span of highway is built by a new operator (as in the recent case of the Brebemi in Italy) the toll is set to cover the full cost of that span; on the contrary, the cost of investments undertaken by an operator who already manages a large network (like Italy's Autostrade) is spread among all users of that network[2]: there is neither economic rationality nor equity in this system, which may have distortional effects on traffic flows. In Italy, we have 24 concessionaires and tolls vary from 4.6 to 14 euro cents per kilometre. These problems are common to countries where tolls are collected, like Spain (see Professor Germa Bel's chapter).

A remedy could be to differentiate tolls paid by users from tolls cashed by concessionaires through a system of taxes and subsidies, which would make evident the tax nature of tolls. But concessionaires prefer another 'remedy', i.e. to obtain extensions of their concessions while maintain existing tolls, with exponential effects on profits, as Italy's history well demonstrates. Extension of concessions is one of the least transparent aspects of highways regulation, facilitated as it is by the fact that consumers do not perceive any increase of costs to them.

Financial cost pricing tends also to cause distortions in investment allocation at country level. Conventions allow operators to recover quickly the full cost of new investments by increasing average tolls over 'their' network with no risk, and public authorities easily approve investments that require no public funding. This tends to cause overinvestment in highways, while investments in normal roads are cut due to lack of public funds. New investments are directed where they can be easily financed through tolls, not where they would be most urgently needed to reduce congestion costs.

EFFICIENCY

Project financing is usually advocated also on the basis of efficiency: licensees would be more efficient than public road departments, and private concessionaires more efficient than government owned ones. The question of whether private companies are in general more efficient has been the subject of innumerable empirical studies, with conflicting results. For companies operating in natural monopolies, the conclusion seems to be that efficiency depends much more upon the quality of regulation than on the type of ownership (Newberry, 1999; OECD, 1997; Willner, 2003).

In the specific case of motorways, there does not seem to be much room for differential efficiency in service and maintenance: technologies are standard, and the risk of over employment to please politicians is limited, as employees' costs are anyhow a relatively small portion of total costs and revenues. In the case of Italy, Autostrade's operating costs declined somewhat following its privatization, but this reflected the reduction of collection costs due to electronic metering, a process that had already started years before: no conclusive evidence can be drawn from this example.

For new investments, one would actually expect private concessionaires to have incentives to save costs and administer construction contracts efficiently, more so than a public road department. But the difference may be minor, if the concessionaire is subject to the same environmental and administrative constraints that slow down public projects, and is bound to assign construction through a publicly regulated tender, like a public road department would do: any difference will depend on how the tender is specified.

Perhaps construction delays could be shortened if the concessionaire was free to choose the construction company he or she pleases. However, in such a case the cost of new investments to be reflected in tariffs is negotiated ex ante between the regulatory authority (ANAS, in Italy) and the concessionaire. Logic suggests that negotiated investment costs tend to be higher than those which would result from a public tender open to all construction companies, and consequently users will have to bear higher tariffs, even if actual costs may ex post be lower. There is also the risk that concessionaires choose construction companies economically related to them even if not the most efficient, if they succeed to obtain higher tariffs to cover the higher costs.

Another point is timing and transaction costs. It takes much longer to choose a concessionaire than a construction company. If a new road is publicly funded, construction contracts may be readily assigned through tenders; in the case of project financing, instead, many other aspects must be considered in the choice of the concessionaire: financial terms, penalties, tariffs' level and their adjustment over time, maintenance obligations, etc. A relevant part of total cost, often estimated at over 10%, is absorbed by transaction costs: no wonder that lawyers and investment bankers love project financing!

A major point in favour of concessions is to reduce the risk that new projects be started with insufficient funding, and then left uncompleted or stalled for lack of funds or because the political patron has changed. Also, concessions make evident risks, administrative and procurement costs which are often hidden in public financing regimes. Overall, the balance of advantages is uncertain, depending upon the quality of the public road department.

PRIVATE OR PUBLIC?

Ministers of the Treasury favour project financing because it is a substitute for public funding. Toll financing is actually a way to introduce a new tax at a low political cost. What are the advantages of granting concessions to private operators, instead of charging tolls by a state agency?

Given the costs of information, contracting and bargaining, state ownership may well be preferred to private ownership (Sappington & Stiglitz, 1987), considering that, in this sector, technologies are standard and the potential advantage of greater efficiency by private operators is very limited.

There may be some competition among private operators only in the rare cases when a concession expires or a concession for building a new road is put on tender. This causes however fragmentation of the system among different operators, with drawbacks already mentioned. There is otherwise no competition among operators: each 'owns' a separate natural monopoly. The role of securing efficiency and preventing extra profits rests entirely on regulation.

Concessions must be awarded for very long periods, often up to 40 years. It is impossible to define precisely contract agreements to regulate tariffs, maintenance, investments for such a long period of time (Kaplow & Shavell, 1999). Regulatory authorities (RAs) have inevitably a wide discretionary power; for instance, if traffic volume risk is borne by the operator, his profit/loss depends largely upon the traffic projected by the RA, when it sets the tariff level.[3]

The potential advantages of a regulation based on price cap are few, given the limited scope for efficiency gains, while the price cap may open the door to extra profits for the concessionaire, if the RA has a benevolent approach in the way it measures productivity gains or quality improvements. If tariffs, under a price cap regime, are revised at short intervals (few years) and the 'claw back' of extra profits is applied at the end of each interval, the system comes to be almost equivalent to a rate of return regulation: the problem remains that of determining which level of profitability should be assured to the licensee (Boitani & Petretto, 1997). If tariffs are revised after long intervals, or the 'claw back' is not applied (as in the case of Italy), the risk of extra profits becomes very large. Tariff regulation based on rate of return seems thus overall preferable to price cap, although, obviously, it further weakens arguments in favour of privatization (see Ogus, 2003; Weyman-Jones, 2003).

Concessionaires have stronger means of pressure than road users, and they may well succeed to 'capture' the RA, also because tariff adjustment may hardly be a transparent exercise, if based on regulatory accounting. Assigning regulation to independent authorities rather than to ministerial offices might reduce such risk (Laffont, 1999). However, as we see in Europe, politicians are very keen to keep control over concessionaires in their hands rather than passing it to independent authorities, presumably because there is much potential return in dealing with companies whose profits depend entirely on tariff adjustments. In Italy, all concessionaires record huge profits, and some have been the 'stars' of the stock exchange for several years. The same happens in Spain or France: enormous financial fortunes are accumulated thanks to highways tolls!

For these many reasons, if tolls are applied, it appears preferable that the operator be a company owned by the government: extra profits would thus benefit the state budget and it would be much easier to impose a socially optimal regulation for tariffs as well as investments (Shleifer, 1998).

The financial costs of raising capital for a government owned company is certainly less than what is demanded by private operators, who require a high return on equity and high premiums for generally low risks (see Professor Sawyer's chapter on the experience of PFI in the UK).

The financial advantage of private funding is largely overstated for at least two reasons. First, a public company in this sector may be highly leveraged, given the stability of revenues; reimbursement of debt may be guaranteed, if needed, by lengthening the period of the concession. Indeed, private capital would not be available for investment if forecast revenues were not amply enough to cover debt service: if revenues are insufficient, government subsidies are anyhow required to launch the project. Second, if the company draws its income from tolls it is excluded from the public sector accounts relevant for the European Monetary Union, even if it is wholly owned by the government.

A public company in charge of roads may have only a limited staff, dedicated to planning and contracting; construction, maintenance, toll collection, services may be separated in many lots and contracted out through tenders, thus fully exploiting the benefits of competition among many suppliers. 'Unbundling' is a system recommended also by the World Bank (Trujillo, 1997).

Unbundling has also another distinctive advantage: it allows a public authority to set tolls for various tracks at socially optimal levels, irrespective of each track's historic construction cost, while maintaining a balance between total network revenues and costs, if this is deemed to be a policy objective.

ROAD TRAFFIC TAXATION

Even granting that a state company would be preferable to many private concessionaires, are tolls preferable to taxes, to cover motorways costs?

In the case of public financing (i.e. if the state covers all costs out of general tax revenues) the net social benefit of an infrastructure (road), over a definite time span (year), is:

$$B(Q) - \mathrm{CT}(1 + \mathrm{dw}) \tag{1}$$

where Q is the traffic flow, $B(Q)$ is the total social benefit (net of private and environmental costs), $\mathrm{CT} = K + cQ$ is the total cost, K being the (constant) financial amortization rate and c the cost of operation and maintenance per unit of traffic flow, and dw is the 'deadweight' cost of tax collection.

If the road is instead financed by charging to users a toll p such that it just covers all costs ($pQ = \mathrm{CT}$), the yearly financial cost becomes $\mathrm{CT} = K + cQ(p) + sQ(p)$, where s is the toll collection cost per unit of traffic flow. Due to the toll, traffic declines from $Q1$ to $Q2$. Financing through tolls is preferable to public funding if:

$$\mathrm{dw}(K + cQ1) > [\Delta B(Q1 - Q2) - c(Q1 - Q2)] + sQ2 \tag{2}$$

The deadweight cost of general taxation must be weighed against the cost of collecting tolls (which for European highways is estimated to absorb 10–15% of revenues; collection stations add about 10% to investment costs),[4] plus the social loss (net of maintenance costs) due to the decrease of total traffic caused by the toll (which obviously depends upon the elasticity of demand).

But highways are also part of a network offering alternative free roads. If highways are free (publicly funded), traffic will distribute itself over the whole network so as to minimize private costs (vehicle costs and time) – as well as social costs (considering only congestion but not other externalities). If instead the use of highways is made more expensive due to the toll, some traffic will divert from the highway to the free road thus increasing private (and social) costs for the users of free roads (due to greater congestion) up to the point where private costs of both are equalized, at an average level higher than in the case of tax financing.

With toll financing there are thus two components of social loss, in addition to collection costs: lower total traffic and diversion of traffic to roads where private costs are higher.[5]

The deadweight cost of tax financing depends upon which tax is considered. If roads are thought to be financed out of fuel taxes (in Italy, expenditures for state roads is only about one-fifth of revenues from transport fuels), collection costs are minimal (certainly much lower than for tolls) and there is no distortion in traffic flows.[6] An argument in favour of tolls based on the avoidance of tax deadweight costs thus seems inconsistent.

In reality, at least in Italy, tolls are not mostly used to pay for the cost of highways, but are just another tax on road traffic, additional to fuel tax. Out of a total 4.7 billion euros paid by users in 2003, some 30% went directly into taxes (VAT, concession tax, concessionaires' income tax). For the largest and oldest concessionaires, operating and amortization costs absorb little over one-third of tolls paid by users. Maintaining existing tolls (and raising them over time) even when investments have been amortized amounts to imposition of a hidden tax on highways users. This is what happened when IRI, the Italian government-owned holding company, privatized Autostrade in 1999–2000[7]: the price paid by private investors (equivalent to 7.7 billion euros for 100% of the capital) was the present value of the tax component of tolls (net of VAT) which the state sold them, for a period of 40 years (Ragazzi, 2004).

Tolls are clearly another tax on road traffic; in Italy they approximately amount to doubling the fuel tax per kilometre, for cars using highways, depending on the vehicle.

Road traffic is heavily taxed in Europe. In Italy, tax revenues from transport fuel, including VAT, amount to over 35 billion euros, the vehicles property tax yields some 4 billion euros (and many other taxes are levied on vehicles), while ANAS spends less than 2 billion annually for investment and maintenance of state roads. Any relation between taxation and cost of roads is purely theoretical.

Fuel taxes may be justified by other reasons (as a consumption tax, to cover accident costs and environmental damages, to limit traffic congestion). No effort seems however directed to determine what is collected for what, and whether the overall level of taxation is socially optimal. Prices of traffic and transport systems are heavily manipulated by governments: What are the costs of the consequent distortions in traffic flows? Are heavy subsidies granted to railways justified? Do we invest too little on roads and too much on railways?

TAXING FOR CONGESTION

To insure optimal use of a given road network, road users should be charged the marginal costs for road maintenance, externalities (environmental and accidents costs) and congestion, so as to restrict traffic from *T* to *T** (Fig. 1). This criterion may also guide investment policy: if there are constant returns to scale in expanding road capacity, then an optimally designed road network has capacity such that congestion charges exactly cover costs of capital and maintenance (Newberry, 1989).[8]

If vehicles, roads and road use were all homogenous, fuel taxes would be the most appropriate tool to levy the optimal charge on traffic. Taxes on fuel are also equitable: the longer, faster and bigger car you drive the more you pay.

Costs of travel are instead extremely variable: certain roads are much more congested than others, congestion varies greatly over time and seasons, externalities vary according to the area, maintenance costs depend upon the type of vehicle, etc. Optimal social use calls for different levels of taxation, which obviously cannot be obtained through a fuel tax: this is, in

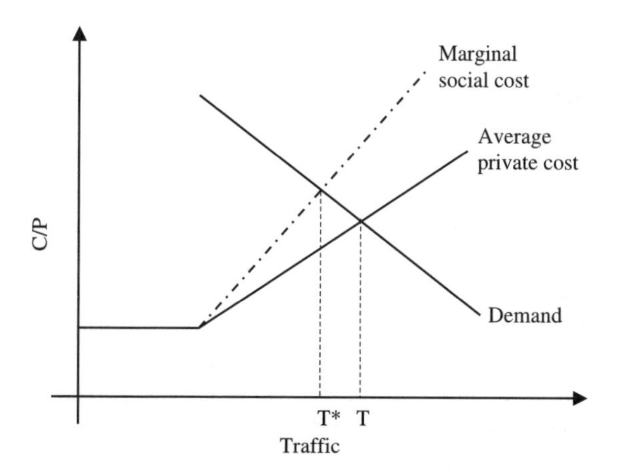

Fig. 1. Pricing at Marginal–Average Cost.

my opinion, the main/only reason that justifies the use of tolls, and it has nothing to do with recovering financial costs.

Focusing only on congestion,[9] differences may (and should) be corrected, up to a point, by building (enlarging) roads where congestion is higher than average. But, even outside cities, it is inevitable that certain roads (bridges or tunnels) have no congestion while others are highly congested.

New technologies are now making it feasible for sophisticated applications to charge different tolls depending on the vehicle, route or time of travel. Various experiments to toll for congestion in cities, as in the case of London, have demonstrated that substantial social benefits may be obtained.

A general rule could be devised to apply tolls according to the degree of congestion (even by the hour) – and, if feasible, also to the degree of environmental damages – where congestion is higher than an average socially accepted level. This would greatly help to quantify and make users and public authorities aware of congestion costs; it would be a powerful incentive to optimal road use[10] and investment allocation. Obviously, no difference is justified between highways and 'normal' roads: congestion tolls should be equally applied to both (if technology allows to do it), and we would perhaps discover that congestion, and the need/ usefulness of tolls, is much greater in some 'normal' roads than on many highways, as the case of Italy suggests.

The UK government is studying the introduction of a nationwide system of tolls related to congestion, revenues from which would be used to reduce taxes on road fuels. Indeed, the wider is the use of congestion tolls, the lower is the optimal level of fuel taxation.

Acceptance of this system ('congestion tolls') by road users would be facilitated if toll revenues were earmarked for specific purposes: in the case of cities, to subsidized public transport, outside cities to build/enlarge the congested road where the toll is levied (Oberholzer-Gee and Weck-Hannemann, 2002). It would help to determine the overall amount to be spent for the road system on the basis of an agreed socially acceptable level of congestion, it would make much more difficult political decisions to build roads in areas where congestion is below average and it would make more difficult for local communities to oppose construction/enlargement of roads since congestion costs are made evident and quantified.

Congestion tolls may be criticized because they discriminate against the poor (all pay the same toll and benefit from similar reductions in travel time, but value attached to time is different). People who attach a high value to time gain from congestion tolling and these are presumably 'rich', although not necessarily so. Empirical studies on this point would be very useful, but I doubt that the adverse distributive aspect would turn out to be sizeable, considering also that revenues from congestion tolls could be used to the benefit of the 'poor', i.e. to improve public transport.

Be as it may, even recognizing that toll pricing for congestion would increase total private costs for some of the road users, it appears socially preferable that part of this higher private cost go into paying for tolls rather than wasted in queues (while air pollution is also reduced).[11]

Tolls are also an appropriate instrument to charge for the differential damage to roads caused by different classes of vehicles, which cannot be recovered through fuel taxes. This justifies tolls on trucks, like those introduced in Switzerland or in Germany.[12] Such tolls may also provide an effective incentive for greater efficiency in the transport industry and for improving the quality of vehicles.[13]

CONCLUSIONS

Financing highways trough tolls (set at a level to cover operating and financial costs) is not justified on equity grounds, while it may cause sub-optimal use of the road system and distortions in investments allocation. Assigning concessions through tenders results in fragmentation of the network and irrational pricing. Each operator runs a separate natural monopoly: there is no real room for competition, and the role of securing efficiency and preventing extra profits rests entirely on regulation.

The difficulty of setting contractual rules for long periods of time, and the risk that the regulatory authority be captured by the concessionaire outweigh the limited scope for greater efficiency or benefits of private funding: if tolls are applied, it is preferable to have one public company rather than several private concessionaires.

Competition could best be exploited through unbundling, i.e. contracting out through tenders the provision of separate services (collection, safety, maintenance of various tracks, etc.)

Old taxes on transport fuels are the most appropriate way to finance road construction and maintenance, including highways. Tolls are, instead, an appropriate instrument to apply differential charges, to bring them closer to marginal social costs of traffic, depending on track, hour of travel, type of vehicle, etc.

Congestion tolls should be applied both on highways and on ordinary roads, and should obviously be administered by a public authority. Acceptance of such tolls would be easier if revenues were earmarked for specific purposes to reduce traffic congestion.

Tax revenues on vehicles and transport fuels largely exceed the cost of roads; much could be gained if taxes, subsidies and investments among different transport sectors were brought closer to marginal social costs.

NOTES

1. MSC for road traffic includes, in addition to vehicle costs, cost of time, road maintenance, accidents and environmental costs, and congestion. Congestion is a reciprocal externality increasing travellers' direct costs (time). Estimates for the US may be found in Levinson and Gillen (1998); for the UK in Newberry (1998).

2. If traffic is not enough to cover costs of new investments, concessionaires obtain subsidies by the government (see the 'Italian case'). Owners of large networks have thus a competitive advantage: they can more easily finance new investments without subsidies, because they can spread the cost over the entire network. However, groups tend to maintain 'their' network managed by separate companies, in order to obtain more easily government subsidies for investments by 'small' companies unable to cover costs with their revenues, without offsetting these financial needs with profits from their other companies.

3. In the case of Italy's Autostrade, in the period 1997–2002 traffic increased twice faster than it had been forecast when tariffs were set for that period; profits were more than double.

4. Time lost queueing at stations should be added to collection costs.

5. The social costs of toll financing may be limited only if alternative free roads do not exist or are too costly due to length or congestion, because in such case demand for highway may be expected to be inelastic and traffic diversion small. The best way to validate toll financing to cover highways costs is to have a very poor network of state roads!

6. With fuel taxes, total traffic is also lower than if roads were financed out of income taxes. However, the decline of traffic, considered above a social loss because we do not consider externalities, is lower than with tolls as average private costs are lower.

7. IRI had originally invested in the company little more than its original capital of 10 billion liras (5 million euros at the conversion rate)! Autostrade's investments had historically been financed by debt, which was reimbursed over time with toll revenues.

8. The environmental damage of road traffic is not limited to air pollution. Our territory is overcrowded; opening a new road has social costs independent form the volume of traffic, which should be added to building costs. Countries where such social costs are higher (Europe vis-à-vis USA) should accept a higher average congestion level and thus also higher fuel taxes.

9. The European Commission (White Paper, 2001) estimates that road traffic congestion costs annually 0.5% of GNP, and that this will double by 2010: congestion is a major social problem.

10. The social benefit of charging for marginal congestion costs depends upon the elasticity of demand. If demand is inelastic, quantitative or technical restrictions may be preferable to pricing policies to control externality costs; see Ponti (2000).

11. Congestion may be reduced also through subsidies: for instance, forms of subsidy (e.g. special lanes, exemption from tolls) could be justified to encourage car pooling.

12. The Swiss system coherently applies tolls according to the distance (and class of vehicle), no matter whether the truck runs on a highway or on a 'normal road'. In Germany, on the other hand, since trucks are charged a toll only on highways, there is the risk of distorting traffic flows towards an excessive use of normal roads.

13. In the case of small countries where cross-border truck traffic is relevant, like Switzerland or Austria, tolls on trucks are justified both to make foreigners pay for costs of roads and to limit environmental damages.

REFERENCES

Boitani, A., & Petretto, A. (1997). Le politiche di regolamentazione. In: A. Ninni & F. Silva (Eds), *La politica industriale: teoria ed esperienze*. Bari: Laterza.

Kaplow, L., & Shavell, S. (1999). *Economic analysis of law*. NBER Working Papers Series, 1.

Laffont, J. J. (1999). Political economy, information and incentives. *European Economic Review*, *43*, 649–669.

Levinson, D. M., & Gillen, D. (1998). The full cost of intercity highway transportation. *Transport Research*, *3*(4), 207–223.

Newberry, D. M. (1989). Cost recovery from optimally designed roads. *Economica*, *56*, 165–185.

Newberry, D. M. (1998). Road user charges in Britain. *The Economic Journal*, *98*, 161–176.

Newberry, D. M. (1999). *Privatization, restructuring and regulation of network utilities*. Cambridge, MA: MIT Press.

Oberholzer-Gee, F., & Weck-Hannemann, H. (2002). Pricing road use: politico-economic and fairness considerations. *Transport Research, part D*, 357–371.

OECD. (1997). *Privatization of utilities and infrastructure: Methods and constraints*. Brussels: OECD.

Ogus, A. (2003). Comparing regulatory systems. In: D. Parker & D. Saal (Eds), *International handbook on privatization*. Cheltenham, UK: Edward Elgar.

Ponti, M. (2000). I costi esterni del trasporto e le linee politiche che ne derivano. *Economia Pubblica*, *5*, 83–98.

Ragazzi, G. (2004). Politiche per la regolamentazione del settore autostradale e il finanziamento delle infrastrutture. *Economia Pubblica*, *4*, 5–33.

Sappington, D. E., & Stiglitz, J. E. (1987). Privatization, information and incentives. *The Journal of Policy Analysis and Management*, 567–582.

Shleifer, A. (1998). State versus private ownership. *Journal of Economic Perspective*, *XII*, 133–150.

Trujillo, J. (1997). Infrastructure financing with unbundled mechanisms. In: Inter American Development Bank (Ed.), *Alternatives to traditional BOTs for financing infrastructure projects*. Washington, DC: Inter American Development Bank.

Weyman-Jones, T. (2003). Regulating prices and profits. In: D. Parker & D. Saal (Eds), *International handbook on privatisation*. Cheltenham, UK: Edward Elgar.

Willner, J. (2003). Privatization: a sceptical analysis. In: D. Parker & D. Saal (Eds), *International handbook on privatization*. Cheltenham, UK: Edward Elgar.

APPLYING A PRICE CAP: RAB AND REGULATORY ACCOUNTING

Pippo Ranci

FOREWORD

The present contribution by an energy expert to a conference on highways should not be misunderstood. I have tried to draw from my experience in energy some ideas which may apply to other utilities. The reader will judge how applicable they are.

ON THE PRICE CAP IN GENERAL

A 'price cap' is nothing but a cap on prices. In the context of regulation it implies that the regulator does not set the price, but he leaves some room for a pricing policy designed and implemented by the company, subject to limitations in the interest of the consumer.

This makes sense, since the service is never perfectly homogeneous. A detailed exercise in price setting by the regulator is a useless show of bureaucratic arrogance. The company may better understand the various needs of the various categories of customers and provide a choice of prices (or of price–quality combinations) while respecting the cap.

A price cap may allow for some flexibility but not much: you always have a maximum price. A regulator can set a cap on the 'average' price of a unit

Procurement and Financing of Motorways in Europe
Research in Transportation Economics, Volume 15, 55–66

ISSN: 0739-8859/doi:10.1016/S0739-8859(05)15005-7

of service, so that the company can offer different combinations between a fixed charge and a unit service charge, or apply different prices at different hours of the day or on different days of the year. A cap on the average price has to be accompanied by strong monitoring to prevent undue discrimination among customers. The company will be wise to apply higher prices to peak-time demand, and this is also socially rational; but applying higher prices to the more rigid sections of the demand curve (Ramsey pricing) may be efficient from the point of view of allocation but unacceptable from the point of view of distribution.

Another possibility is to set a cap on revenues. When capital cost is the main component of the cost of service, a fair rate of return can be provided by a sufficient level of overall revenues, regardless of the unit price of the service sold. So the regulator is inclined to calculate and set a revenue cap: if demand increases, the company will have to lower the unit price. This way, the company has no incentive to increase the volume of sales: this can be good or bad according to whether you consider sales promotion by the company as socially positive or negative. On environmental grounds it is widely held that promotion of energy consumption is not socially desirable: it follows that a revenue cap is preferable to a price cap in the case of electricity or gas. In the case of highways the reverse may be preferable: under a price cap the company is pushed towards increasing the utilisation of the network, so relieving the other roads.

Setting a cap on total revenues, or on average prices (equal to average revenues), implies a choice of a level of generality: the regulator can impose a cap on the (total or average) revenues from selling the service to a category of customers, or a general cap on all revenues. This implies different degrees of possible cross subsidies among categories of customers. The trade-off is between risking to allow cross subsidies and setting too rigid a discipline, which would prevent the company from meeting different tastes or needs of different types of customers, or from practising a rational peak-load pricing policy.

There is a different notion of the price cap, dealing with adjustment in time. This has to do with the task of designing price controls which include an incentive to efficiency: in other words, the task of artificially reproducing the incentives, which are usually provided by competition, in a context of regulated monopoly. Here comes the well-known CPI-X (Each year the tariff is increased by a factor proportional to the Consumer Price Index (CPI) and decreased by a factor measuring the increase in efficiency (X) that the regulator expects from the company, on the basis of experience and comparisons) formula.

So we have a static and a dynamic notion of price cap. Both can be expressed in terms of capping the price (tariff) or the revenue. Both have an importance in regulation, but have different roles. In any debate it is important that it be made clear which one is being discussed.

A NOTE ON ITALY

According to the handbook approach, a tariff-setting procedure based on the notion of price cap is introduced with the purpose of introducing incentives to efficiency which the traditional cost-of-service (or rate-of-return) methods did not provide.

This is not the whole story, and the Italian case may provide an example of a wider scope for the price cap.

Price-cap-type reforms were introduced into the Italian tariff picture in the early 1990s (tariffs for highways and for water services) and then developed in the new legislation on independent regulatory authorities (Law No. 481/1995). Tariff reform was seen as a way of reducing political discretion and introducing a stable, independent technical regulatory approach. This had to do with a widespread dissatisfaction with the traditional approach to tariff setting by government offices, which was too discretional and too undependable.

The setting up of independent regulatory bodies was seen as one chapter in a wider reform of the institutions, in the direction of increasing the stability of institutions and rules, and decreasing the interference by company lobbies and political parties.

At the same time, a fundamental choice was made in favour of liberalisation of industries previously dominated by legal monopolies, and of privatisation of state-owned enterprises. Even before actual privatisation, the state-owned enterprises started to behave like private companies: the government asked the managers to 'create value', in view of increasing the contribution of these companies to the state budget (both via dividends and via proceeds from the sale of shares). The same happened to municipal enterprises, equally set for future privatisation.

In the new context a stable frame of rules, including price regulation, was essential to the operating of all companies and to the securing of adequate investment decisions: no longer could the future supply of public services (which implies a timely investment in new infrastructures and plants) be guaranteed by direct government command on public enterprises; rather, security of supply would be provided by a stable frame of fair rules which would induce companies to invest.

So the price cap method became a symbol of a transparent, reliable way of setting prices for services not (yet) open to competition, and a flag of a new approach in protecting the consumer and the public interest in view of a wider scope for private initiative in some strategic sectors of the economy.

DEFINITION OF COSTS

A price cap does not differ from a cost-of-service or rate-of-return rule, in that they all depend on having a good estimate of the cost of producing and supplying the service. The same difficulties apply.

I will briefly illustrate two issues.

First, a choice should be made with respect to the definition of costs: average costs? Or marginal costs.

The common choice is in favour of average costs, which are easier to assess and are directly linked to the profitability of the enterprise. A basic principle of a fair price setting by a regulator is that prices should generate enough revenue for the firm to cover all direct costs and depreciation, and allow a fair remuneration of invested capital. This is easily done by calculating the necessary level of total and average revenues.

Reference to long-run marginal costs (LRMC) is theoretically preferable. In practice, the LRMC rule provides a significantly different outcome, with respect to an AC rule, only if the cost of new capital is different from the cost of existing capital, and it can be estimated with sufficient approximation. If the life of capital assets is very long and the cost of new capital is higher than the cost of existing capital, an LRMC rule will create a rent in favour of the company, which may be difficult to justify.

A different case for an LRMC rule is that of tariffs for peak load, or for the management of congestions. Usually, these are special cases, to be dealt with through special regulation, or by setting a broad revenue cap and leaving it to the company to find the appropriate tariff structure. Unfortunately, there is still scant experience with such economically rational tariff-setting procedures.

The second issue is more crucial. The traditional methods imply setting the tariff at a level where all actual costs incurred by the company are covered; no incentive to cost reduction is provided. On the contrary, modern tariff setting implies setting the tariff where standard costs of a normally efficient company are covered, leaving an extra profit to highly efficient companies and allowing the inefficient companies to run a loss; if the level of recognised costs is known in advance, companies will have a strong incentive to reach as high a level of efficiency as possible. So, in order to set the price cap, standard costs to be recognised should be determined in advance.

Although rational, such incentive-compatible price regulation is rarely understood. Companies whose costs are not covered will strive to show that this is not a consequence of their own inefficiency but of some external factors such as some characters of the territory, or the composition of types of customers, or specific public service obligations, or other circumstances. When the cap is set, they may appeal against it, and the court may decide that a company providing a public service has a right to have enough revenues to cover all costs actually borne. As an example, such decisions have been taken by administrative courts in front of regulatory decisions on gas tariffs.

COLLECTING THE DATA

Any method of price setting necessarily requires detailed information on costs. The regulator relies on information provided by regulated companies; a good company accounting is essential.

The crucial step is the unbundling of activities. When the exercise began in the mid-1990s in Italy, many municipal companies had not yet separated the accounts of electricity from those of gas and sometimes other services; Snam provided overall costs for the gas service where it was not possible to separate the cost of imported gas from the cost of national gas and the cost of transport. So the problem is not an outright manipulation of data by the companies, as sometimes stated in the handbooks, but information insufficient for the purposes of regulation.

If there is one main company the regulator has to rely on it, as in the case of gas imported by Eni-Snam; if the company refuses to provide the information required, a lengthy legal controversy may arise. Once the regulator reaches a good understanding of the costs, he has trouble in moving from actual to standard costs, i.e. costs not inflated by inefficiencies which can be removed. One can look at companies abroad and adopt average parameters such as the number of employees per unit of product, but such an exercise is open to criticism and the results of its use can easily be appealed. So at the end of the day the most advisable procedure is: (a) ascertain the costs as provided by the company, (b) make an estimate of the (lower) level of costs of an imaginary efficient company of the same size, (c) allow a reasonable number of years for the company to reach the latter cost level starting from the former, and (d) set the initial tariffs at the level of the actual costs and introduce a parameter of required or expected productivity increase (an X parameter) such that the recognised costs will be lowered to the imaginary cost level of an efficient company in the chosen number of years.

If the companies are many and small, as in the distribution of gas, a large collection of data is necessary. The obstacle is the poor level of accounting in many small companies.

Standard costs may be estimated by calculating the average level of costs in a relatively more efficient subset of companies.

The whole exercise is very delicate and companies will fight to avoid any reduction of historical tariffs. Even when all the data on costs indicate that the traditional tariffs are too high, any reduction will raise public complaints and possibly even legal appeals. The regulator must adopt very transparent procedures and explain what he is doing in the clearest way: this is no guarantee of not being challenged, but it helps.

COMPONENTS OF RECOGNISED COST: THE OPERATING COSTS

Operating costs may vary over time, while the character of a price cap is that it is set for a number of years, normally 4 or 5, so that the company can rely on fairly sure forecasts of future revenues on the one hand, and has an incentive to reduce costs on the other.

The costs of labour and of intermediate inputs are subject to some increase if the rate of inflation is positive; since the price cap includes adaptation for inflation, the company will strive to maintain labour and other cost increases below the allowed inflation rate.

Some costs cannot be properly foreseen: a typical example is the cost of fuels in electricity generation, which depends on the world price of oil.

Here, a special indexation procedure is required. It can be so designed as to respect the characters of the price cap, i.e. certainty and incentive to efficiency: the indexation mechanism has to be transparent so that any cost increase can be traced to its origin, and it must be automatic so that the company that buys fuel for the future at a time when the cost is low can make a profit.

COMPONENTS OF RECOGNISED COST: THE CAPITAL COSTS

The cost of capital is the product of an estimate of the assets to be remunerated and the rate of return on such assets.

Setting the value of the assets for regulatory purposes (the so-called Regulatory Asset Base or RAB) does not imply that the same value of the assets has to be used for all purposes. Of course, the easy way is to use the company accounts and calculate industry averages on the basis of standard capital requirements per unit of service provided.

Two decisions have to be taken: the standard amount of recognised capital per unit of service provided and the unit value of such capital. Once a standard capital intensity has been set, companies employing higher than average quantities of capital will complain and try to show that this is not a consequence of their own inefficiency but a necessity given by the specific environment in which they operate.

But the main difficulty lies in assessing the unit value of the assets. Here, the choice between historic and replacement costs is debated. Replacement costs are theoretically preferable but it has to be recognised that most of the networks will never be replaced; they will require more or less radical maintenance operations but in general the sites will be maintained, so it would be inappropriate to remunerate the assets by using today's cost of land. Historical costs, although adjusted for inflation, may be preferable.

This opens the way to recognising a different value to new assets: if such a provision is not introduced, a reinforcement of the networks which is socially desirable may not be decided by the companies. In the old regime, and in many utility sectors, new investment was decided by government planning; now we rely on company decisions, and we have to set the appropriate incentives to make sure that all socially useful developments are decided on and built.

THE RATE OF RETURN ON ASSETS

Which is a 'fair' rate of return? Basically, the financial markets should provide the answer. Tariffs should include a remuneration of capital that is no more and no less than what is necessary for the company to raise capital on the market.

The standard procedure is to calculate a 'weighted average cost of capital' (WACC).

Since equity capital costs more than borrowed capital, the weights to be included in the WACC procedure make a difference. As usual, standard industry weights are preferable to the actual company weights: a company can pursue its own strategy in setting its capital mix, having future goals in mind, but the tariff level should not be influenced by company strategies.

In setting the rate of return for both equity and borrowed capital, the standard procedure is to set a risk-free rate of return first, and then add an industry risk component and possibly a country risk component. The difficulty comes from the insufficient development of the financial markets: utility companies whose shares and bonds are traded on the stock exchanges are few and present large differences in the industry and country mix.

All rates of return should be defined as real, i.e. net of inflation, since the tariff is then adjusted for inflation (see the following paragraph). If a nominal rate of return is chosen as a component of the tariff, and then the whole tariff is indexed, we have a double counting of inflation.

COSTS AND TARIFFS IN TIME

A standard adjustment procedure follows a multiplying formula $(1+\text{PI}-X+Q)$.

A price index (PI) has to be chosen. The traditional choice of the consumer's price index, which has also been adopted by the Italian legislator, does not reflect the changes in the costs of the inputs. According to the industry, a different price index has to be selected. In the case of energy (electricity and gas) the indexation mechanism is very important, since fuels are a main component of costs, so the cost of fuels has been separately indexed to the world prices of a basket of fuels, and the rest of the tariff follows a general inflation index.

Where the cost of fixed capital is a sizeable component of the total cost, and physical assets have a long life, such as in the case of highways, a question arises: should the indexation process be applied to the whole tariff, including depreciation? This may appear to provide an excessive remuneration in the case of old assets which may have been financed at low, pre-inflation, interest rates; on the other hand, if real interest rates are used in setting the return on capital initially (see preceding paragraph), it is fair that inflation is taken into consideration when adjusting the tariff. It is not unfair that a good or lucky choice of finance may increase the company's profits, and a bad or unlucky choice may produce losses for the company and not be charged on consumers.

The productivity factor (X) should reflect the average expected gains of (total factor) productivity in the industry. In practice, in newly liberalised utilities the initial level of efficiency can be quite low, due to an excessive number of employees who cannot be fired abruptly. In such cases, the X factor includes a gradual approach to standard productivity levels; here, the size of the parameter does not depend on estimates of efficiency gains produced by technical progress, but, more simply, a discretional decision on the length of the adjustment period, which reflects a compromise between the search for efficiency and the social constraints. Such a decision includes an estimate of social costs, and a regulator is uneasy in taking it; yet, if the regulator refuses the responsibility of such a decision, the outcome may be even less rational.

The natural increase in demand produces benefits through an increased exploitation of the existing fixed capital; this is particularly true in the highly capital intensive industries where capital utilisation is usually far below saturation, such as highways. In such cases, a price cap (as distinguished from a revenue cap) will produce an automatic increase in profits year after year, unless the growth of demand is included in the X parameter. So the X parameter in a price cap regime should be higher than in a revenue cap regime, ceteris paribus.

The adjustment formula, and indeed the tariff discipline, should be set for a predetermined length of time: the so-called period of regulation, usually 3–5 years. A long period provides the benefit of a greater certainty for investors, and the disadvantage of increasing the effects of any error in estimation or forecasting, or of any external, unforeseen development.

Should we apply the X parameter to the whole tariff? It has been argued that a gain of productivity can be expected in the use of labour or of other inputs, much less so in the use of existing capital; if the argument is accepted, then the X factor should not be applied to depreciation and to the cost of capital. But I have doubts on this argument. Gains in efficiency can lead to a better use of any input in the production process, although in different proportions, so that any gain in efficiency is a weighted average of improvements in the use of the various inputs, including capital. I am ready to support this stand in the case of a revenue cap, although with some doubts; I have no doubts in the case of a price cap.

Productivity increases always produce an increase in profits, or a reduction in losses; a price cap type of tariff setting creates an incentive for the managers to introduce productivity gains because the company accounts will benefit from them, irrespective of the level of the X factor. Different levels of the X factor adopted will affect the company results, but they will not alter the incentive.

The incentive greatly depends on the length of the period of regulation. Investment completed in the middle years of the period of regulation will start producing benefits exactly when the regulator collects data on costs, in order to set the tariff level for the next period; so the regulator will observe the fall in costs and set lower tariffs, and the benefit of investment will be transferred to the consumers very quickly. Therefore, the company has an incentive to concentrate the investment decisions in the extreme years of the period of regulation, when the benefits will be kept in the company for longer. Or the company can obtain a carry-over provision from the regulator: a share of the productivity gains attained in excess of the predetermined X path will be maintained even in the following period of regulation (the Italian government has imposed such a clause on the regulator, to the benefit of the state-owned energy companies). Of course such a provision interacts with the tariff-setting process that will be followed for the next period of regulation, and it can be offset by it.

Tariff can be adjusted for changes in the quality of service. The regulator can establish a link between the present tariff and the present quality levels, and while designing an adjustment path for the tariff in time (the X factor), he may introduce a desired path of improvement in the level of quality. Then, the tariff will be raised if actual quality turns out to be better than the predetermined path, and lowered in the opposite case. This is a rather sophisticated instrument, and it implies different tariffs for different providers of the same service. It has been applied in few countries, while a different discipline has been introduced by the Italian energy regulator, which is consistent with maintaining a common national tariff.

CROSS SUBSIDIES

In principle, tariffs should reflect costs and cross subsidies should be avoided. Cross subsidies have been heavily used in the past, as an instrument of industrial, regional and social policy. It is one of the main tasks of modern regulation to restore a clear connection between costs and prices, and eliminate distortions. But it is not easy, since each distortion has a historical origin providing an appearance of rationality, and is supported by a lobby, sometimes a very powerful one.

Yet, some cross subsidies are inevitable, and you can find them even in the free market whenever a company sets a price for a product which applies to units sold in different places and under different conditions.

When setting a tariff for use of a network, a basic problem has to be solved: should the tariff be set for a unit of the network used (say, a kilometre of line or pipe or highway) or should it be set for a unit of service (say, a postage-stamp tariff for an envelope of mail, or a kilowatt-hour of electricity sent, irrespective of the point of origin and the point of destination)?

The postage-stamp approach is not based on social considerations alone: it may have some merits even on the ground of allocation theory, since an electricity grid works like a system of water channels, where water can be poured into here and taken out there without the molecules of water actually travelling all the way. In fact, some transactions run counter-flow and hence contribute to the capacity of the network rather than using it.

So the choice of a tariff system is not obvious. The gas transportation tariff is based on an 'entry–exit' method, where the effect of each operation on the use of the network is estimated with the help of a simulation model. Such a method is applied nation-wide, so its rationality weakens when dealing with cross-border trade.

I am not familiar with highways: I imagine that cross subsidies at the company level are quite normal, although not quite rational; at a regional or national level, a highways tariff system should take some account of externalities, and this raises the issue of cross subsidies again.

LIBERALISE TARIFFS?

Tariffs are not set for ever: in most cases they are necessary during a transition to a competitive liberalised market in which free prices can be trusted to be fair to customers.

In general, this does not apply to the network tariffs: it is a common belief that the networks remain a natural monopoly. This is the case for the electricity and gas networks, even if in some cases a competition from merchant lines can be envisaged.

On the contrary, competition among networks is quite normal in telecommunications. Here, technical advances are unique and the physical constraint of relying on scarce land has been overcome: the case cannot be generalised.

So in any sector the possibility of competition among networks has to be practically assessed. It will not be enough to have some threat by alternative networks; effective competition by a rather large number of reciprocally independent companies is required before allowing free prices. This is not a likely outcome in most sectors.

TRANS-EUROPEAN NETWORKS: EU INFRASTRUCTURE PROPOSALS

John Hugh Rees*

INTRODUCTION

European transport policy has shown a remarkable legislative development in the last 15 years. However, this has been accompanied by a rise in problems, notably in relation to the environment and accidents, that cast doubt on the sustainability of the transport system as it currently exists. For this reason European transport policy must be conceived as a long-term action that is consistent with the strategies for economic growth (Lisbon) and for durability (Gothenburg). This is far from an easy task and the role of infrastructure is crucial as this chapter aims to illustrate.

THE POLICY IN THE EUROPEAN UNION AND TRANSPORT

It is widely agreed that energy and transport are key elements to ensure economic growth and cohesion. However, the key role of European

*Corresponding author. Mr Rees is currently an adviser in the Direction General for Energy and Transport at the European Commission in Brussels. The views expressed in this paper are personal and do not represent the position of the Commission.

Procurement and Financing of Motorways in Europe
Research in Transportation Economics, Volume 15, 67–73

ISSN: 0739-8859/doi:10.1016/S0739-8859(05)15006-9

transport policy in EU development was apparently not sufficient to ensure that it was initially given an important place in Community policy. For years a real 'Community approach' to transport was quite simply ignored. The Member States maintained their control over the development of the transport sector. As previous efforts had not proved sufficiently fruitful, the new Commission (2000) decided early in its mandate to give a fresh impetus to transport and hence the White Paper of 2001.

The challenges that faced the Union at the beginning of the new millennium were quite clear:

- First, congestion was increasing and becoming a serious problem on many main roads, around major cities and at principal airports. The cost of congestion was estimated to be 0.5% of the total national wealth in the EU.
- Second, transport was making an increasingly important contribution to environmental problems. Nearly a third of CO_2 emissions come from transport – 80% from road transport.
- Third, the number of accidents, particularly on the roads, was unacceptable – about 50,000 people are killed every year in the enlarged Union.
- Fourth, the dependence of the transport sector on one source of fuel – oil – exposes the sector to more and more risks in the event of problems in supply and increased prices; 97% of transport depends on oil for fuel.

The Commission has a limited number of policy 'levers' available for its use:

- first, legal acts (regulations, directives, etc.) that are based on the various articles of the Treaty;
- second, economic measures – also based on the Treaty – but aimed at changing the rates of taxes or charges;
- third, grants to assist projects, such as transport infrastructure, having an important structural goal;
- fourth, demonstration programmes to promote new technologies or best practices that can sometimes be accompanied by financial assistance; and
- fifth, research work to develop and promote new ideas – such as the use of hydrogen in transport.

The Commission has to keep very close to the line taken in the Treaty. Transport policy – a common policy like agriculture – is set out in a separate Treaty Title. But, unfortunately, the transport Title (V of the current Treaty) is not really very clear as it tried to maintain a balance between the fundamentally open market principles of the Treaty in general and the special or social objectives that concern major parts of its transport sector. This is important as proposals have to try to achieve a fine balance between economic efficiency and wider, social objectives. The aim has to be to try to ensure that transport meets the needs of industry and citizens while at the same time becoming more environment friendly and, in time, sustainable.

The White Paper accepted that in the time period up to 2010 the best that could be achieved would be to prevent the situation from getting worse and start the process to create a truly efficient, sustainable system.

A particular problem that concerns all the Members States is the shortfall in public investment in transport infrastructure. The growth of road traffic has been due to the increase in the number of private cars and the development of road haulage thanks to the liberalisation of the market in the 1990s. From Fig. 1 it can be seen that the rate of growth that the transport sector has been achieving is higher than that of the economy as a whole, as transport has become relatively cheaper. At the same time the governments have been putting less money into transport. At the beginning of the 1980s most Member States invested the equivalent of 1.5–2.0% of their GDP on transport infrastructure. In the 1990s and through to today this figure has come down to 1% of GDP.

The consequences of this reduction in investment in new capacity in a period of high traffic growth are clear – congestion.

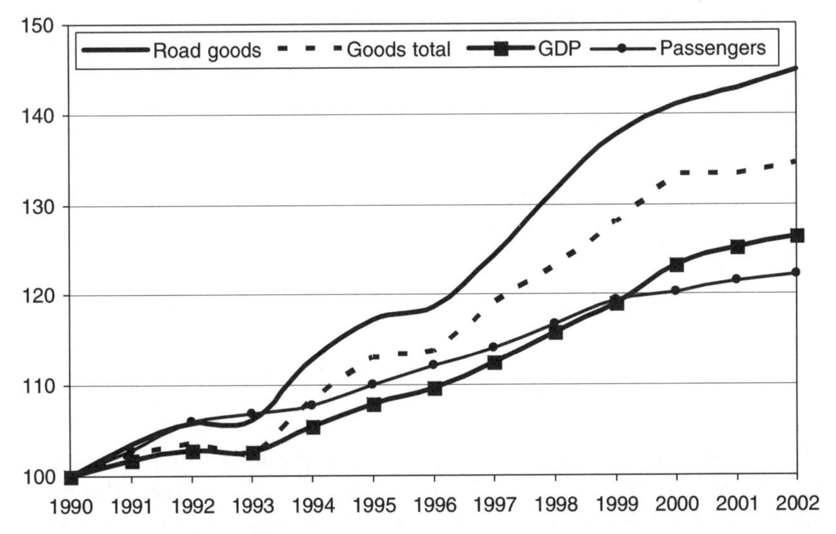

Fig. 1. The Development of Transport in the EU 15.

To tackle this the White Paper, while it proposed an expansion of investment in the trans-European networks, recognised reality and recommended that better use is made of the spare capacity, which does not require heavy investment to be brought into service. Basically, this means sea and inland waterways plus the railways. To achieve this objective, these modes will have to become far more efficient and attractive to customers. The White Paper also stressed the importance of the so-called 'inter-modal' transport.

Environmental sustainability would also be at least partly tackled by making a greater use of water and railways, which are more environment friendly than road transport. However, road transport is and will remain the most important land mode but there should be more priority given to making clean vehicles and fuels – renewable if possible – and re-thinking of the whole taxation system for transport fuels. Actually existing taxes are an inefficient way of paying for the use of transport as they are unable to differentiate between, say, rural areas and areas in cities with heavy congestion. The White Paper recommended that a new approach to infrastructure charging be applied to all modes that would take environmental or external costs into account and could differentiate between different locations, and types of vehicle and time.

On the roads the Commission has posed the ambitious target of halving the number of deaths (40,000 p.a.) by 2010 – this can be done because the difference between the best and the worst Member States is a factor of 1:2 – achieving the standards of the best states everyone would achieve the desired results.

THE PRINCIPAL MEASURES PROPOSED BY THE 2001 WHITE PAPER

In total, the White Paper proposed over 60 separate measures concerning all modes of transport.

If these measures were approved, the overall result should be to return to the modal share of transport in 1998 by 2010. What does that mean in practice?

For many years virtually all the growth in the land freight market has been met by increased road transport. For passengers it is the same picture with the growth shared between road on the shorter distances and air for longer trips. If the modal shares of 1998 are to be restored by 2010, this implies that there should be a slowdown in the growth rates of road and air transport with the additional growth being absorbed by other modes, notably maritime and rail. In terms of figures this would mean that road freight transport would increase only by 38% in the period – as compared with the almost 50% increase without the measures. This growth should be compared to the expected growth of the economy of 43%. For passengers the increase in passenger car movement would be limited to 21%. The growth is picked up by the water transport and railways, which have to take about 40% more traffic by the end of the period.

THE NEED FOR INFRASTRUCTURE

New infrastructure is called for even if the overall objective is to control the increase in demand for transport services. The concept of the trans-European networks was launched 10 years ago at the Maastricht European summit but it has never fully taken off because of problems in securing an adequate budget. As has already been noted this failure, when linked to the fact that the Member States have cut back radically on their investments in transport infrastructure, is the reason for many of the bottlenecks found even today on the road system in particular. However, the Community has recently (2004) re-launched the networks. It was agreed in April 2004 that there should be 30 priority European projects that receive the bulk of the EU financial resources for transport. To undertake these 30 projects by 2020 is estimated to call for around €225 billion. The Commission believes that this concentration of resources is the best way forward but an appropriate budget is needed. Set against the €225 billion of priority projects cost the current budget of €600 million is really too small even if, in some countries, the networks are supported by other Community operations like the Cohesion fund. To tackle this problem the Commission has asked for a fourfold increase in resources in the new budgetary period starting in 2007. If this larger budget is agreed on Community aid can have a real leverage effect if it is combined with additional resources coming from the private sector – this is the concept of a 'public–private' partnership that the Commission wants to exploit to the maximum. The Commission has made a detailed assessment of future financial requirements under the new perspectives for the period 2007–2013. Just to be clear how the €20 billion figure has been arrived at let me explain that this assumes that the 30 projects will require around €140 million in the period and that the EU budget will produce 15% of that, i.e. €20 billion. Again, to clarify, I should recall that these 30 projects were agreed on with a detailed timetable by the European Parliament and the Council of Ministers and for the Commission it is entirely logical to expect that this decision is now followed up by an appropriate decision on financing. In practical terms the financing of the sections of these projects which involve frontier crossings is from past experience always the most difficult. For this reason the Commission has proposed that such sections can receive a higher level of support than the rest of the project.

This Community support would be largely used to exert a 'leverage' effect on the financial markets to create a degree of interest in the provision of loans. In addition existing users of routes in the neighbourhood of the major projects would be called upon, through pricing systems, to make a contribution. This approach is rather similar to the technique of 'cash flow' financing that has been used with success for major projects in the past. But, of course, pricing is important not only for financing new projects but also to create the right conditions to regulate the use of all modes of transport infrastructure. In order to increase transparency in infrastructure financing and justify the charges levied, there has to be a clear link between real costs and the charges. In this way the charging system overall can be used to ensure that the infrastructure system is used rationally and in the best interest of the economy and society. As for the Community the two issues of particular importance are ensuring that prices are equitable and do not discriminate between users and that the prices are set at an appropriate level to ensure that external costs are met. The progress made with the taxation of heavy goods vehicles (the 'Eurovignette') shows that the EU is starting to move in the right direction.

FUTURE DEVELOPMENT

What progress has been made? In terms of legislation we have advanced quickly – this can be seen from Fig. 2.

There are several areas that have already been identified where further efforts are called for:

- The first area concerns sustainable development where the indications are that the transport sector is not going to meet its objectives under KYOTO. It is probable that the issue of external costs will have to be looked at seriously and a comprehensive review of pricing for all modes undertaken. This should include the aviation sector, where de-taxing kerosene has no justification other than it has been that way for a long time. More efforts will also have to be made in promoting clean vehicles and fuels.

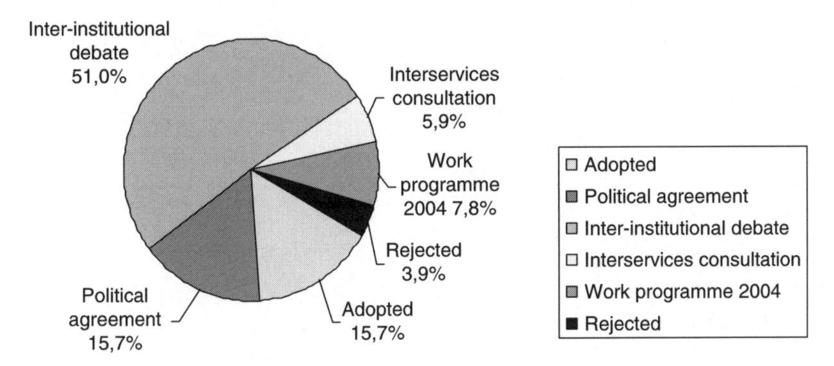

Fig. 2. Progress with Proposals in White Paper 2001.

- In this context the Community has been absent from the urban transport field and in view of its importance I believe thought should be given to measures that will help the major cities get to grips with their traffic and environmental problems.
- Further action to extend the progress on passengers' rights in the aviation sector to other modes is also called for. Perhaps this could also be extended to certain forms of goods transport.
- Security has not been mentioned but it is certain it will be a key issue in the coming years and the transport sector, as we have seen lately in Madrid, is a clear target for action.

In the course of 2005 the Commission is committed to produce an interim review of the 2001 White Paper. The review should summarise the results achieved to date and indicate any new subjects to engage and areas where further action is called for. Overall, it is clearly too soon to expect that many of the measures, which have only recently been approved, will be producing identifiable results. Some targets, like road accidents, show indications of improvement, but in others, e.g. railway modal show, there are no indications as yet. The EU has succeeded to a large extent in introducing new legislation but it will take time to see whether this legislation will overall produce the deserved results. As far as transport infrastructure is concerned the outcome of the deliberations on the new budget in the Council and in the European Parliament will be absolutely crucial in achieving the objective of creating a system that meets the EU's growth requirements in a sustainable manner.

MOTORWAYS AND MOTORWAY FINANCE IN GERMANY AND AUSTRIA

Werner Rothengatter

INTRODUCTION

In Germany as well as in Austria the motorways were constructed through state activity from the beginning. While in Austria the paradigm of public provision, management and finance has changed since 1982, when the State-owned ASFINAG[1] company under private law was founded, in Germany there is still the traditional public regime for the motorways. The road users pay taxes, but these are in effect not hypothecated for road infrastructure but instead go into the general budget. Presently only one third of the total tax payments of motorists consisting of fuel and vehicle taxes as well as the VAT on the fuel tax are spent on road infrastructure. This has led to serious complaints from the industry and other stakeholders.

Against the background of static or even decreasing tax income and the rising costs of infrastructure provision, a change of the financing paradigm is being discussed in Germany, greatly stimulated by a high-level government commission on infrastructure finance, which submitted its suggestions to the government in 2000. The commission, named after its chairman the Pällmann Commission (2000), argued that a change from tax- to user charge-based finance was at the same time inevitable and efficient in the

Procurement and Financing of Motorways in Europe
Research in Transportation Economics, Volume 15, 75–91

ISSN: 0739-8859/doi:10.1016/S0739-8859(05)15007-0

economic sense: inevitable because of the deteriorating financial situation of the public budget, facing the international race towards lower income taxation, and economically efficient because of the built-in incentives to use the infrastructure in a better way.

Adding to that, the European Commission have strongly contributed to encourage the member countries to introduce motorway charging at least for heavy goods vehicles (HGV). A white paper on fair and efficient pricing for the use of transport infrastructure initiated in 1998 the discussion on flexible motorway charging in the "Eurovignette" countries (Benelux, Sweden, Denmark, Germany and Austria), which had introduced this instrument of time-based access charging for HGV since Jan. 1995. In 1999 directive 1999/62 EC was introduced for charging HGV on motorways in the Union. The White Paper of 2001 on common transport policy underlined again that the user charging on Trans-European Networks is a necessary instrument to harmonised competition in the transport sector.

In Germany the government planned to start with an electronic kilometre-based charge on HGV on motorways in September 2003, while Austria intended to follow in January 2004. The German plan failed because of serious problems with the charging technology, whereas the technically more robust electronic system used in Austria (developed by the Italian Autostrade) has started according to plan.

In this chapter the development towards this change of the financial paradigm and its consequences for the public budget will be discussed. The principles of charging and the institutional settings which are associated with the change of paradigm will also be the focus of analysis. As the present concepts of motorway charging are partial approaches, which are associated with a number of failures and shortcomings, the chapter will conclude with future prospects towards a more complete economic concept of network-wide charging of road traffic.

HISTORY OF MOTORWAY CONSTRUCTION IN GERMANY AND AUSTRIA

Motorways are roads with at least two lanes in one direction and free of level crossings. A first road of this type was built in 1909 in Berlin, the so-called AVUS (Automobil-Verkehrs- und Uebungs-Straße: road for automobile traffic and training). In 1926 a study society was established to develop a national network of motorways, and an association was formed for the construction of a motorway from Hamburg via Frankfurt to Basel (Hafraba e.V.), which came out with first plans in 1933.

While in Italy the first – privately financed and tolled – Autostrade was constructed since 1923 and the extension stopped after 1935, in Germany a most intensive phase of motorway construction was started in the mid-1930s. The motorway link Frankfurt–Darmstadt was opened on 19 May 1935, and in 1936 altogether 26 new sections followed. In 1942 a network of 2,128 km existed, which interconnected the major agglomerations in Germany. Construction activity was extended to Pomerania, which after World War II became Polish. Altogether, 405 km of motorways have been built in the now Polish territory between 1936 and 1942. Parts of the tracks are still existing but have not been rehabilitated under the socialist regime.

After the annexation of Austria in the year 1938 this country was integrated into the network extension; the plans had already been prepared earlier. The motorway Munich–Salzburg had already been completed and the construction of the motorway Salzburg–Vienna was started in March 1938. Short sections were opened at Salzburg until 1941 but then the motorway construction work was stopped because of the increasing war activity. The length of the Austrian motorway network was 16 km only, after World War II. In 1954 the construction work on the corridor Salzburg–Vienna (about 300 km) was continued. North–South connections and other links followed and extended the network to 1,670 km in 2003 (see Fig. 1).

In Germany the motorway construction was continued in 1953. About 50% of the West German network had been constructed by 1975. Today this network comprises about 12,000 km.The present configuration of the German motorway network is shown in Fig. 2.

HISTORY OF ROAD FINANCE AND CONTRIBUTIONS OF ROAD USERS TO THE PUBLIC BUDGET

Germany

Provision of road infrastructure including the motorways was and is still regarded as a public issue in Germany. Earmarking is in general not possible, according to the budget law, but was applied in some special cases. One of these cases is the earmarking of the revenues from additional fuel taxation after a tax increase in the 1960s and early 1970s to finance the road

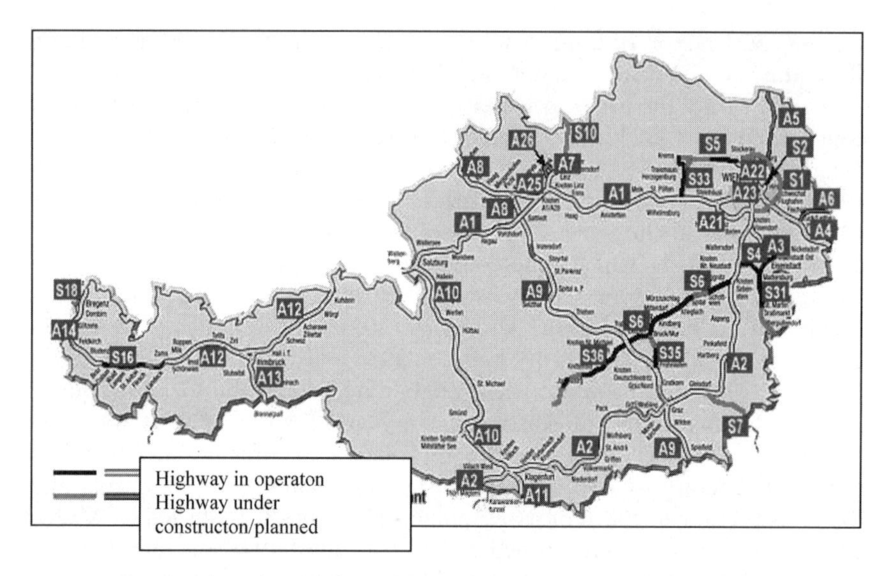

Fig. 1. Austrian Highway Network (Motorways Double Lined).

extension. The financial law of 1971, extended by the municipal funding law, allocated fixed parts of the fuel tax to federal roads (25%), local roads and local public transport. While the municipal funding law was transposed according to plan, the earmarking of revenues from fuel taxation for federal roads was stopped already in 1973 by a simple legal trick. The financial law of 1971 is set out of force by the yearly budget law since that time.

In 1999 the ecological tax was added to the fuel tax, which is earmarked for social insurance to keep labour costs low. The idea was to get a double dividend from this strategy. First, the fuel consumption in transport would drop with positive effects on the environment and second, the reduced labour costs would stimulate employment. It is estimated that the wage share allocated to social insurance would have increased to 22.5% in 2004 while with the help of the ecological tax it could be lowered to 19.3%. Green groups argue that this reduction has saved 80,000 jobs. Presently, the tax load on fuel sums up to €654.50 and €470.40 per 1000 l for gasoline and diesel, respectively (for lead- and sulphur-free fuel). The overall income was €43 billion in 2003.

Fuel and ecological taxation go into the federal budget. Also, the VAT on fuel taxation (16%) goes into the federal budget, but is partly redistributed on the basis of fixed shares to the states and communities. Vehicle taxation

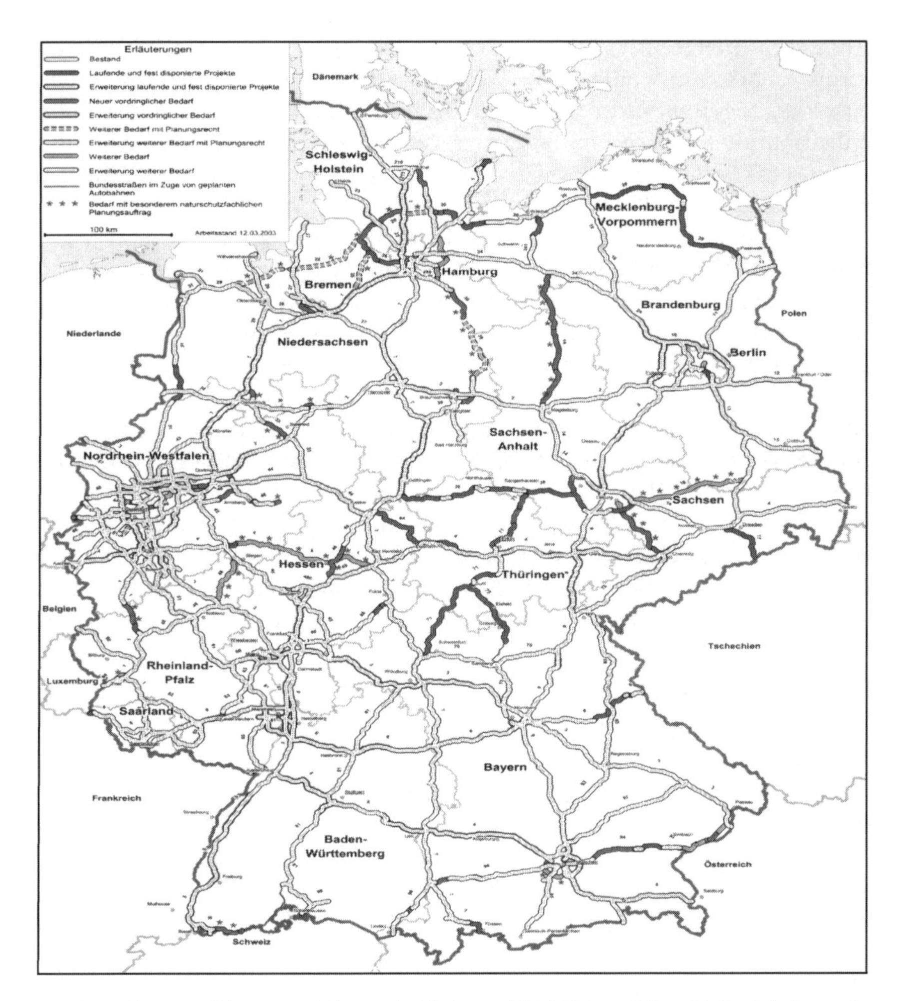

Fig. 2. German Motorway Network. Coloured Full Lines: New Links (Pink, under Construction/Decided; Red, High Priority; Yellow, Further Needs). Coloured Double Lines: Extensions of Existing Motorways (Pink, Red, Yellow as Above). Green Dots: Further Treatment Subject to Special Environmental Risk Analysis.

goes into the budget of the states (Laender). The revenues are not earmarked. This difference in financial competence related to the taxation of transport is relevant insofar as it often has been recommended to substitute the (yearly fixed) vehicle taxation through increased fuel taxation to strengthen the incentive effects of the (variable) fuel taxes. Such a restructuring of taxation would be very complicated in Germany because of the necessary negotiations between the federal government and the states on the redistribution of tax income. The total vehicle tax income was €7.6 billion in the year 2003. The yearly tax income from the transport sector is exhibited in the statistics of DIW (2004).

Summing up the revenues from fuel and vehicle taxation one arrives at a figure of roughly €50 billion for 2003. Estimating the expenditures for roads at the federal, state and community levels at about €16 billion, one can conclude that only one third of the road-related tax income is spent on roads again. This low share of tax income allocation to the road sector, which is the source of the tax money, has been questioned by industrial associations and other stakeholders, who argue that motorists pay too high taxes. The counter-arguments are that (i) the expenditures do not reflect the real costs, which are much higher, (ii) public overhead costs are substantial, which are not included in the infrastructure expenditure figures, (iii) external costs should be considered, which in several studies are estimated to be much higher than the infrastructure costs, and (iv) taxation has to be shifted from income to indirect taxes because of the global pressure to lower income and corporate taxation.

Austria

Austria changed the financial regime for the federal road and rail infrastructure in 1982. The ASFINAG was established for planning, constructing, maintaining, managing and financing the motorways and highways, i.e. a network of about 2000 km length. ASFINAG is a company under private law, which is completely owned by the federal state. In 1997, the company received further tasks and the right to charge tolls and fees for the utilisation of motor- and highways.

ASFINAG is spending around €120 million per year for the maintenance of the network and about the same sum is allocated to renovation/rehabilitation (re-investment). The investment for new construction and extensions was between €400 million and €600 million in the last 3 years. These expenditures are fully financed by tolls. The Austrian toll system consists of three components:

1. Since 1997 all vehicles up to 3.5 tonnes have to buy a vignette;
2. Since January 2004 all vehicles with a gross weight of more than 3.5 tonnes pay a kilometre-based toll, which is collected electronically (see

section titled 'Principles of charging and Charging Schemes' for more details); and

3. For five routes (Pyhrn, Tauern, Brenner, Arlberg and Karawanken) additional tolls are charged. Further Alpine passes are considered for additional tolling.

Point (3) is not completely in line with EU legacy and the Commission is trying to move Austria to abandon the double charging of motorways. By revising the Directive 1999/62 for charging HGV the Commission tries to offer alpine countries the opportunity to raise higher charges in environmentally sensitive areas so that the double charging would become unnecessary. However, Austria is fighting for HGV charges, which are high enough to control the development of transit freight traffic on roads. The first drafts of the revised Directive gave not enough flexibility to apply an ecologically based demand management, which is a reason for Austria to oppose for a long time.

PLANS FOR THE FUTURE DEVELOPMENT OF THE MOTORWAY NETWORKS AND THEIR FINANCE

Germany

Federal Transport Infrastructure Investment Plan

The development and rehabilitation of the federal road network is part of a federal transport infrastructure plan (BVWP), which includes road, rail and inland waterways. As airports (with minor exemptions) are not owned by the federal state, airports are not included. But the access links to airports are part of the BVWP. The BVWP procedure starts with a forecasting of all influencing factors of traffic, in particular population, the national and regional economy and the international changes. The traffic forecasting is based on this background and presumes a 'default' capacity extension of the transport infrastructure. The suggested projects (about 1,500) are assessed by means of a monetary benefit–cost analysis, an environmental risk analysis and a spatial development analysis. European interconnectivity, synergetic and substitution effects as well as intermodal effects among the different modes of transport are taken into account.

Altogether the financial budget for federal roads is €28.9 billion for current projects and €49.8 billion for new projects until 2015. It includes

€14.3 billion for 1,600 km of new motorways and €12.6 billion for 2,250 km of extended motorways (increased number of lanes).

Financial Instruments
Without going into all the details of budget management one can classify the financial instrument into five categories:

1. Funds from the federal budget;
2. European funds (TEN, ERDF, Cohesion);
3. The programme "Investment for the Future" (2001–2003), financed by income from tendering the Universal Mobile Telecommunication System (UMTS) licenses;
4. Anti-congestion programme (2003–2007), financed by the expected income from HGV charging on motorways; and
5. Public–private partnerships in the form of "F- and A-models".

The programme "Investment for the Future" was launched after the tendering of UMTS licences to telecom companies in the year 2000. The income was altogether about €50 billion, which has been spent on the reduction of public debt and some investments in education and transport. The contribution to the BVWP is €2.3 billion. The anti-congestion programme should be funded from the revenues stemming from the motorway charging on HDV, which the Ministry planned to introduce in September 2003. However, because of technical difficulties, the introduction of the charging system had to be postponed to January 2005 so that a financial gap has emerged in the meantime. As the federal government has brought this case to court it is not clear to what extent the gap will be closed by penalty fees of the Toll Collect consortium or by public credits.

Public–private partnership has been in vogue in Germany since 1994, when a law was passed which allows for private investment in roads[2]. This law opens the possibility for private investors to construct, operate and finance roads of a similar standard like motorways (not motorways themselves), tunnels, bridges and alpine passes. The so-called F-model sets the framework for such a regime and allows for charging HGV, light goods vehicles and cars, based on public price regulation. One project of this type has been realised, which is the Warnow-crossing tunnel in the city of Rostock, where the main investor is the French company Bouyges. The tunnel was opened in September 2003 and, unfortunately, the financial success is disappointing so far. This influences the further promotion of five other projects, which are considered for procurement according to the F-model. As the project list comprises only 31.2 km of new investment it is obvious that the F-model cannot contribute substantially to the finance of federal roads.

The so-called A-model gives private investors the opportunity to invest in motorway extensions and to finance the venture by the allocated income from HGV charging. As HGV charges contribute, on an average, about 50% of the total costs of motorways, investors can apply for public co-finance in a tender process. In the BVWP, 12 projects are considered for A-model finance, which comprise 522.6 km length with a budget of €2.2 billion.

Austria

Also, in Austria the strategic plan for developing the motorways and federal roads results from a political process supported by standardised assessment studies. The forthcoming revision of the master plan is presently worked out and will be the basis for negotiation with ASFINAG on the priority setting and finance of the planned extensions. In contrast to the German case, the process of promotion and procurement is much more influenced by the professional company, while in Germany the project proposals are brought forward by political actors.

As the federal roads in Austria have to be financed by tolls, every proposal for new investment has to be checked by ASFINAG with respect to its financial viability. This creates a natural barrier against over-investment, which is missing in Germany. Although the green movement is strong in both countries it seems that the political activity in Austria is more focussed on controlling the environmental impacts of road transport. Ecological regulations of trucking, high motorway charges and even higher tolls for alpine passes and tunnels give – together with priority investments for the railways – a better chance of influencing the modal split in favour of environmentally more friendly modes. A basic element of this policy is the electronic charging on HGV with 3.5 tonnes gross weight or more.

PRINCIPLES OF CHARGING AND CHARGING SCHEMES

Preliminaries

In both countries there are two leading objectives for charging road traffic:

1. Infrastructure finance and commercial management; and

2. Traffic management to achieve goals of efficiency, equity and environmental protection.

While the financing issues are predominant in these countries, they also try to integrate environmental aspects into the pricing schemes. This is done through appropriate differentiation of prices and by spending the revenues on projects which support environmental sustainability. In the following we will focus on charging HGV.

Existing Legacy: Directive 1999/62 EG and Planned Revision

European legacy sets the basic rules for pricing HGV $\geqslant$12 tonnes of gross weight on European motorways. In the case that traffic safety on roads other than motorways is negatively affected, such roads can be included in the pricing scheme to avoid traffic diversion. Directive 1999/62 EC defines first that only cost elements, which are directly related to the provision and operation of road infrastructure, are the basis of cost-oriented charges, i.e. external costs are excluded. Second, the charge for a vehicle category has to be based on the average infrastructure cost which is allocated to this category. Third, it is possible to differentiate the charges according to two criteria: the time of the day (peak/off-peak) when the maximum difference between the highest and the lowest price does not exceed 100% and the environmental performance, measured by the Euro standard for vehicle emissions, while the maximum difference between the highest (for high-emission vehicles) and the lowest (for low-emission vehicles) price does not exceed 50%. The German government has decided only to make use of the second method of price differentiation, i.e. to differentiate according to environmental performance.

For about 2 years the EU Commission is working on a revision of Directive 1999/62 EC in the following directions:

1. Extension of the priced road network (all Trans-European roads and secondary roads in case of potential traffic diversion);
2. Reduction of weight limit from 12 to 3.5 tonnes;
3. Earmarking of revenues;
4. Consideration of environmental sensitivity (markups for passes through the Alps and the Pyrenees);
5. Consideration of accident costs; and

6. Differentiated guidelines for calculating infrastructure costs, in particular omission of capital costs for old infrastructure.

The problems discussed in the foreground of the negotiations are often influenced by political strategies in the background. For instance, the peripheral countries want to get cheap access for their haulage industry to the transport markets of the countries in the European core, and the countries in the core want to preserve the domestic markets for their domestic transport industries.[3] This is the reason behind the planned differentiation between 'new' and 'old' infrastructure, which does not make any sense from the economic point of view, but seems to be agreed upon by a majority of Member States. The revision has been agreed by the Council in April 2005, while the approval of the Parliament is still open. Not all of the issues listed above have been transposed as it is discussed in more detail in the paper of Borgnolo and Rothengatter.

PRINCIPLES OF CHARGING AND COST ACCOUNT

Directive 1999/62 EC determines the economic basis of charging (see EC, 1995; EC, 1998; EC, 1999). It is allowed to distribute the total cost of HGV to the vehicle categories. The average cost can be varied by congestion and environmental parameters while the overall revenues should not exceed the overall costs. The accounting structure is then characterised by capital costs (depreciation and interest on capital) and running costs. Capital costs represent the largest cost block while their evaluation provides most of the problems associated with the accounting and allocation procedure (see Littlechild & Thompson, 1977; Knieps, Küpper, & Langen, 2000; Prognos, IWW, 2002; Rothengatter, 2003). Summarising capital costs and running costs, one arrives at the total volume of costs to be allocated to the vehicle categories. This cost allocation is performed following the principles of causality, responsibility and fairness, which finally leads to more than 100 segments of cost allocation.

Using figures and forecasts on transport performance, vehicle categories and environmental classification the motorway cost allocation results in the suggestions given in Table 1 for HGV charging in Germany.

The tolls actually charged to HGV in Germany are about 17% lower than those exhibited in Table 1. The reason is that the government had promised to the road haulage industry to pay back a compensation of €600 million in

Table 1. User Charges Differentiated by Axles and Environmental Categories.

Year	No. of Axles	Category A	Category B	Category C
2003	Up to 3	10	13	15
	4 and more	12	15	17
2005	Up to 3	11	14	16
	4 and more	12	16	18
2010	Up to 3	10	12	15
	4 and more	12	15	18

Note: Category A: Euro 0 and 1; Category B: Euro 2 and 3; Category C: Euro 4 and 5.

the form of a tax reduction for fuel taxes. As this plan was not agreed on by the EU Commission, it was decided to cut the tolls until an agreement is achieved on the toll compensation.

TECHNOLOGY OF THE PAYMENT SYSTEM IN GERMANY

The German payment system is organised on three platforms, which follows from the political requirements:

- Manual payment at ticket machines (posted at gasoline stations and other central places);
- Payment through the Internet; and
- Electronic payment through an on-board unit.

The third platform is an innovative element of the payment scheme, developed by the Toll Collect consortium.[4] The Toll Collect electronic payment scheme is based on two telecommunication architectures: GSM radio communication and GPS satellite navigation. Entry to and exit from the motorway network as well as the route chosen is controlled by GPS. The on-board unit calculates the distance in kilometres and the associated charges according to the vehicle type (pre-coded in the On-Board Unit (OBU)). The information collected by the OBU is transmitted via GSM communication to the Toll Collect centre, which sends out the invoices.

According to the contract with the government the Toll Collect consortium is allowed to use a part of the revenues for recovering the cost of installing and operating the payment scheme. If the revenue forecast comes true then this part accounts for 17% of the total revenues, which seems quite high. The consortium argues that additional benefits can be realised by the companies because value-added services can be received through the communication system, as for instance pre-trip, on-trip and post-trip information. The government wants to foster technological innovation and Toll Collect is the first system which fulfils the future requirements for standardised electronic tolling systems in Europe. The future costs of such electronic systems could be much lower because of learning effects or increasing returns to scale and to scope.

CHARGING PRINCIPLES AND CHARGING SCHEME IN AUSTRIA

The charging principles in Austria are similar to those in Germany, which is due to the fact that the advisory teams of these countries held close contacts and tried to avoid unnecessary heterogeneity (see Herry, 2001). Therefore, it is not necessary to go into the details of the costing calculus. The reasons why the Austrian charges [5]are substantially higher than the German are the lower traffic density in Austria and the higher cost of investment in alpine areas (Fig. 3).

The payment regimes and the system of the electronic payment schemes differ considerably.

In Austria all goods vehicles with gross weight from 3.5 tonnes upwards are charged. Vehicle owners can buy a so-called GO-box for a small amount

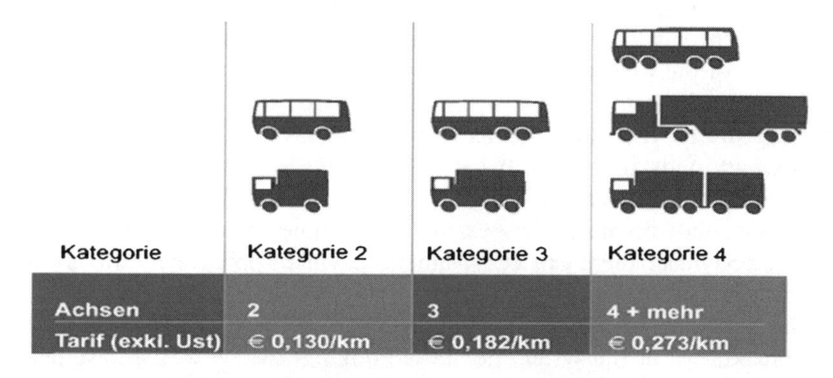

Fig. 3. Tariff Structure on Austrian Motorways.

of money (€5), which can be fixed at the screen shield. The signals for the GO-boxes are generated at gentries, which are located close to the entries/exits of the motorways. Technically, this is done by DSRC, i.e. short distance radio communication. The users can choose between pre- and post-payment after receiving the usual invoices.

EXPERIENCE AND EXPECTED IMPACTS FROM CHARGING ROAD USE

Switzerland and Austria

The Swiss system was started in 2001, and the 3 years of experience allow for a first evaluation of impacts:

- The aim of stabilising freight traffic on roads and transit movements in particular has been achieved. Truck-kilometres went down by 10%. In this context it has to be considered that the maximum gross weight of trucks has been increased from 28 to 34 tonnes (and 40 tonnes since Jan. 2005).
- The financial revenues in the year 2003 are €541 million; the system costs are €38 million, which is about 7% of the revenues.
- Transalpine freight traffic of the railways has increased substantially, i.e. there are positive modal split effects.
- The environmental impact is twofold. First, the reduction of truck-kilometres and the change of modal split have led to reduced emissions. Second, the differentiation of tariffs according to the environmental performance, based on Euro standards of vehicles, has induced a rapid change of the vehicle fleet structure. The emission reductions achieved by the second reaction are much higher compared with the first reaction.
- There are no traffic diversions from motorways to secondary roads because all roads are priced. It has to be noted that diversion effects which could occur through choosing the shortest paths instead of the fastest paths cannot be measured because in the time before starting the electronic payment there existed a vignette payment only for motorway use, which has been abandoned after.
- The electronic payment system works reliably. There has occurred neither a breakdown of the system nor false bookings, which would have led to complaints or trials at court.

The Austrian payment system was introduced in January 2004, so that there are only a few observations, which are not significant in the statistical sense:

- A stabilising effect on freight transport on roads is not reported yet; at least, it seems to be much smaller compared with Switzerland.
- A change of the vehicle fleet structure towards modern, environmentally efficient vehicles seems to be going on, but slowly. In this context it has to be considered that environmental policy towards trucks has been very strict in Austria in the last years so that the haulage companies were stimulated to use new technology (the so-called Eco-point quota policy).
- Traffic diversion from motorways to secondary roads seems to be higher than forecasted. Regional politicians complain about the increased truck transit traffic through cities and villages. This is a result of restricting the charging scheme to motorways only (European law).
- The technical system works reliably; there are no reports on major failures. A first comparison of revenues and costs of the electronic payment system shows that the system costs are below 10% of the total revenues.

Germany

As the electronic payment system could not be introduced in September 2003, as originally planned, there is no practical experiences until now. This means that there are only estimated or forecasted impacts to be derived from a limited number of studies. Reference is made here to a study launched by the German Environmental Agency, which was elaborated by IWW (2002). The Environmental Agency was concerned about the possible diversionary effects of tolling, which can cause counter-productive environmental and safety effects. Four types of reactions were in the focus of the study and led to the following roughly summarised results:

- *Diversion of traffic from the motorways to the secondary network*: On an average about 3% of HGV is diverted to secondary roads, according to transport modelling results. In the environment of agglomerations this share can be higher such that the loads on the secondary networks can grow there by 8–10%;
- *Diversion from road to rail*: Modal effects are low if the service quality of the rail is left unchanged. If the logistic quality of the rail service can be increased there will be a higher intermodal reaction.
- *Strategic adjustments of logistics, roundtrips and loading factors in the road haulage industry*: These reactions will increase with increasing charges. With an average charge of 15 cts/km the effects will be modest.

- *Change of the vehicle fleet*: Because of the differentiation of charges according to Euro-emission categories there will be considerable incentives to buy new trucks with modern technology. The environmental effect stemming from these incentives is much higher than the effect from social marginal cost pricing.

PLANNED INSTITUTIONAL CHANGES AND CONCLUSIONS

In Germany the VIFG company has been established, which will administrate the allocation of revenues from HGV charging on motorways. For the next years the VIFG will be left fully integrated in the public regime, because the revenues from HGV charging will flow to the Ministry of Finance and are transferred to the VIFG afterwards. The VIFG in the present form has no competence for taking credits, planning or managing infrastructure provision. But there is a chance that this company will develop in the future to a more autonomous body. The advantage of the VIFG foundation is that – although this company presently has almost no competence – it can develop systems of monitoring and governance for the F- and A-models of public–private partnership in Germany.

In Austria the basic institutional setting with the foundation of ASFINAG is regarded a success so that only minor changes are to be expected. The company has the feeling that there is still too high an influence of policy in the management activities. Changing the legal form from a stockholding to a limited liability company would enlarge the degree of entrepreneurial flexibility. Therefore, this change of the legal form is at stake presently.

Summing up, the Austrian financial policy in transport is more consequently addressing the issues of infrastructure finance and environmental costs while the German policy, is still dominated by the general public choice approach, controlled by the Ministry of Finance.

NOTES

1. ASFINAG stands for Autobahnen und Schnellstraßen- Finanzierungs-Aktiengesellschaft.
2. Fernstraßenbauprivatfinanzierungsgesetz, 1994; F-änderungsgesetz, 2001.
3. Note that since 1998 cabotage is allowed in the EU. There are only some exceptions for accession countries for a transitory period.

4. Partners of Toll Collect: Deutsche Telekom, DaimlerChrysler and Cofiroute.
5. For HGV $\geqslant$12 tonnes 22°cts/km are charged in A on an average and 15°cts/km in D.

REFERENCES

DIW. (2004). *Verkehr in Zahlen 2003/4 (transport in Figs. 2003/4)*. Berlin: Ministry of Transport, Construction and Housing (BMVBW).

EC. (1995). *Towards fair and efficient pricing in transport policy – Options for internalising the external cost of transport in the European Union.* Brussels: Green Paper of the European Commission, COM(95)691.

EC. (1998). *Fair payment for infrastructure use: a phased approach to a common transport infrastructure charging framework in the EU. Brussels*: White Paper of the European Commission, COM (98)466.

EC. (1999). Directive 1999/62/EC: of the European Parliament and of the Council of 17. June 1999 on the charging of heavy goods vehicles for the use of certain infrastructure. *Official Journal of the European Community, L 187*, 42–50.

Herry, M. (2001). *Wegekostenrechnung fuer Oesterreich.* Vienna: Study Commissioned by the Ministry for Transport and Innovation.

IWW. (2002). *Anforderungen an eine umweltorientierte Schwerverkehrsabgabe (Requirements for an environmentally oriented tolling system for heavy goods vehicles)*. Berlin : Study Commissioned by the German Federal Environmental Agency.

Knieps, G., Küpper, H.U., & Langen, R. (2000). Abschreibungen bei Preisänderungen in stationären und nicht stationären Märkten. In: *Diskussionsbeiträge des Instituts für Verkehrswissenschaft und Regionalpolitik* (Vol. 71). Freiburg.

Littlechild, S. C., & Thompson, G. F. (1977). Thompson: aircraft landing fees: a game theory approach. *The Bell Journal of Economics, 8*, 186–203.

Pällmann Commission. (2000). *Final report of the governmental commission transport infrastructure financing.* Berlin: Minister for Transport, Building and Housing.

Prognos, IWW (2002). *Wegekostenrechnung für das Bundesfernstraßennetz unter Beruecksichtigung der Vorbereitung einer Streckenbezogenen Autobahnbenutzungsgebuehr.* Berlin: German Ministry for Transport, Building and Housing (BMVBW).

Rothengatter, W. (2003). How good is first best? Marginal cost and other pricing principles for user charging in transport. *Transport Policy, 10*, 121–130.

FRENCH MOTORWAYS: EXPERIENCE AND ASSESSMENT☆

Alain Fayard, Francesco Gaeta and Emile Quinet

THE FRENCH ROAD NETWORK

The French road network includes 976,000 km of roads, 38,600 km of which compose the central government's road network (national roads, free and tolled motorways) under the responsibility of the French Roads Directorate. The rest of the network is run either by counties (360,000 km) or by local districts (595,000 km). The French regions (France being divided into 22 regions), which contribute to the financing of the national and district road networks, do not manage any road. It is worth noting that the whole French road network carries roughly 78% of freight traffic and 90% of passenger traffic, the remaining being mostly borne by rail.

On the central government's road network, which corresponds to 4% of the whole road network, 40% of traffic is concentrated, in particular on motorways and more generally on toll motorways (7,840 km), which support a length-weighted average of 26,400 vehicles per day. Generally speaking, the average daily flow per kilometer is about three times larger on motorways than on ordinary national roads and about 30 times larger than on local roads.

☆Views expressed in this chapter are those of the authors and not necessarily those of the organizations which the authors belong to.

Procurement and Financing of Motorways in Europe
Research in Transportation Economics, Volume 15, 93–105

ISSN: 0739-8859/doi:10.1016/S0739-8859(05)15008-2

The intercity motorways, which build the framework of the national network, are usually tolled. The density of the French motorway network is about 14 km per 1,000 km^2 and 140 km per million inhabitants. According to the recent law 2004-809, a significant part of the national road network (20,000 km out of the present 38,600 km) will be transferred to counties from January 2006. Only the core network – the main axes of national or European interest – will remain managed, directly or through concessions, by the national government, which represents approximately 18,000 km including the whole motorways network, 7,840 km of which is under concession. Actually, the road services of the central government are going to be reorganized. At the present time, the French Ministry of Infrastructure and Transport is structured on the basis of a central administration based in Paris and a local network, both at the regional (Direction régionale de l'équipement) and at the county (Direction départementale de l'équipement) levels. In the near future, 11 interregional road directorates (Directions interrégionales des routes) will be set up in order to manage the 10,000 km of trunk roads and not tolled (toll-free) motorways (roughly 5,000 km of ordinary roads, 5,000 km of double carriageways and motorways). The idea is to run national roads on the basis of "routes" superseding the existing area organization. According to each itinerary, the interregional road directorates will provide users with management and traffic information centers along with maintenance and intervention centers, distributed along the network approximately every 60 km. The regional directorates will be in charge of the "maîtrise d'ouvrage "(i.e., "owner's management").

TRAFFIC AND LENGTH

Traffic is highly concentrated on a small part of the network, mainly the motorways and the national network, as mentioned above. It is clear from these figures that the bulk of the traffic and its growth is on the motorways, especially on toll motorways; this evidence stems first from the fact that motorways bear a much larger flow than the other parts of the network, and second from the dramatic development of the motorway network, as shown in Table 1.

To be more precise, at the present time the motorways network is composed as shown in Table 2.

Table 1. Motorway Network Development.

Year	Network Motorway Length (km)	Of Which Toll Motorways (km)
1960	170	10
1970	1,560	1,060
1980	5,010	3,730
1990	6,910	5,515
2004	10,400	7,840

Note: From French Roads Directorate statistics.

Table 2. Structure of Motorways Network.

Motorway network length (km)	10,400
Toll motorways (km)	7,840
Toll free motorways (km)	2,560
2 × 2 lanes to become motorways (km)	1,035

FINANCIAL ASPECTS

The financial aspects of the road sector can be appraised from several points of view. In particular, the allocation of funds devoted to roads, split according to the source (user, national taxpayer, local taxpayer), has changed along the years. The part of concessionaires has steadily increased from 32% in 1975 to 53% at present, while the share of governmental funds has decreased from 56 to 27%, the remaining share being covered by the local authorities, mainly by the regions. At the present time, expenditures carried out by concessionaire companies on 7,840 km of toll motorways are bigger than public investment (governmental and local funds) in the rest of the national road network (30,600 km). Public and private funds are used both for maintenance and investments, the split between these two categories being made as shown in Table 3.

As far as tolls are concerned, French motorway concessionaires have collected €5,832 million in 2003, which means an increase of 5.30% in 2002, while the average daily traffic was 26,426 vehicles per day: cars, 84%; heavy goods vehicles (HGV), 16%. In 2003, the average toll tariff per kilometer (including VAT) added up to 6.87 euro cents for light vehicles and 19.64 euro cents for HGV; HGV traffic generates approximately a third of the toll revenues. According to concession companies' statistics, in recent years,

Table 3. Breakdown of the Expenditures on the French National Road Network in 2003.

	Tolled (7,840 km)	Untolled (30,600 km)	Total
According to the use			
Maintenance	0.7	0.6	1.3
Investment	2.1	1.8	3.9
Total	2.8	2.4	5.2
According to the source			
National Government budget	1.4[a]		
Concessionaires	2.8		
Local authorities	1.0		
Total	5.2		

Note: From French Roads Directorate statistics (in billion €).
[a]Government staff not included (estimated €500 million).

tolls per kilometer have increased by about 1% and toll revenues by about 5%, the difference being due to the increase in traffic.

Toll revenues are used by companies to finance operating costs related to customer services (maintenance, safety, etc.), depreciation, financial costs linked to interest and repayment of loans (i.e., investments in infrastructure), VAT and other taxes. Out of €10 of toll collection, about 40% constitutes depreciation and financial charges, 20% operating costs, 10% net return, the remaining being VAT and other taxes and duties.

According to Decree No. 95-81 of January 24, 1995, the toll tariff evolution in France is fixed on the basis of 5-year planning contracts signed between the Government and each concessionaire. These contracts provide a frame of reference in terms of legal and economic conditions covering a midterm horizon. The concession contracts are the legal basis of the concession, and the 5-year planning contracts allow fine-tuning. In particular, they deal with three main aspects:

- the companies' agreed objectives concerning maintenance and investments in infrastructure and services;
- the companies' agreed objectives concerning road safety, social policy, and environmental protection; and
- the obligations and rights in terms of toll tariff evolution according to the financial and toll policies pursued.

The first generation of 5-year contracts drew to a close at the end of 1999. Because of the ongoing reforms in the concession system and VAT regulation, the preparation of new contracts was held up until 2001. Five new contracts have been signed since then and others will be signed in the near future.

HISTORY OF THE MOTORWAYS NETWORK[1]

1955–1969: A Commitment to Tolls with Public Companies

In the 1950s the main post-World War II rebuilding work was over and car ownership began to increase rapidly. By 1951 the Government established a special dedicated road fund (FSIR), which was to receive a percentage of motor fuel tax receipts, but competing budgetary pressures prevented the Government from funding the FSIR in full. Consequently, in 1955 a law was passed to allow toll financing of motorways ("autoroutes"). Public control was to be maintained by granting concessions only to a local public organization, a chamber of commerce (public body in France), or a "mixed" company in which public interests have a majority of shares. Moreover, the 1955 law stated that the use of "autoroutes" is, in principle, "free" but the exception rapidly became the rule within a decade since, between 1956 and 1963, five "mixed" companies were set up (these companies were called "sociétés d'économie mixte concessionnaires d'autoroutes", or SEMCAs). Nevertheless, the initial concessions were for only short portions of motorways (50–70 km), except in 1963 for the top priority, the south–north axis part between Lille, Paris, and Lyon (130- and 160-km segments). All five SEMCAs shared a similar financial and organizational structure: they were very weakly capitalized (€100,000–€300,000) and the only shareholders were public bodies; the national equity stake was held by the Caisse des Dépots et Consignations (CDC), a State-owned investment bank. The Government provided initial financial assistance by guaranteeing the loans of the SEMCAs and providing cash and advances that were fairly significant (averaging 30–40% of construction costs). Throughout the 1960s the SEMCAs were little more than paper organizations, nothing more than the "false nose of the State" as a minister said:

- A State-owned investment bank (CDC) marketed the State-guaranteed loans for the SEMCAs through a special office established in 1963, Caisse Nationale des Autoroutes, which only pools the borrowing of the SEMCAs.

- Another subsidiary of the CDC, the SCET, managed the accounts and works contracts.
- The Road Directorate in the Department of Public Works and Transport designed and operated the motorways (except toll collection). Private contractors carried out the construction.

1970–1981: Liberalization and Privatization. Emergence of Cross-Subsidies

At the end of the 1960s only 1,125 km of intercity motorways were in service. A reform was set up in order to: (i) allow private companies to compete for new concessions and (ii) strengthen the existing public companies (the so-called SEMCAs) to give them more autonomy and responsibility. Between 1970 and 1973, four private toll road companies were awarded contracts for 300–500-km motorways each (except for a 63-km-long concession). All four new concessionaires were consortia of major French public works companies; no investors were interested in investments with such a long payback period and banks were said to bear shares more willingly because they wanted to support contractors they were linked to and to issue bonds than because they wanted to invest. The Government was less generous with assistance for concessions granted in the 1970s than it had been in the 1960s. Nevertheless, significant financial aid remained available to concessions granted to both private and public companies. For example, in the case of the first private company Cofiroute, 10% of the funds was covered by equity, 10% by in-kind advances from the State, 65% by State-guaranteed loans, and 15% by loans without guarantee, that is, 75% of the funds was issued or backed by the Government. At the same time, the public companies established with SCET a new company Scetauroute to act as their "maître d'oeuvre" (engineer), their engineering firm and prime contractor for construction and research on motorways. The public companies created their own maintenance services. Increasingly, the motorway companies were expected to subsidize new stretches with surpluses generated on their older segments, which had higher traffic and had been built at a lower cost. Moreover, the dates at which the concessions on their older and more lucrative sections expired were often extended. A system of cross-subsidization within companies appeared gradually; it undermined the concept itself of the profitability of the individual motorway segments and even of the company (for public companies). Last but not least, the concession agreement of the four private companies stated that the toll rates would be set by the company in the limits determined in the concession agreements; this procedure was extended to public companies. Nevertheless, in 1975 the

Ministry of Finance declared that it would regulate tolls. Therefore, tolls came back firmly under the Ministry of Finance's control and this Ministry can, in this way, control the entire toll motorway system. This breach of contract was backed by the administrative Supreme Court, "Conseil d'Etat", relying on a 1942 price control law (CE No. 01139 01146 01147 01159 May 13, 1977).

1982–1994: Facing the Crisis, a Nationwide Mechanism of Consolidation Network

At the beginning of the 1980s the motorway system faced serious problems of cash deficit, a reason for which (but not the only reason) was the oil crisis. The State took over three of the four private companies and indemnified shareholders, which was a soft enforcement of the forfeiture clause. Moreover, in 1982 the Government established in place of FSIR a new dedicated fund, the FSGT (Special Fund for Public Works). This fund was allowed to issue bonds to give a leverage effect to the additional tax on top of fuel tax, which was earmarked for it. On average, public resources dedicated to road (budget + FSGT) were stable during the period of existence of this Fund until 1987: The FSGT's increasing resources were compensated by decreasing budget funds.

With respect to the public companies, a new government agency, Autoroutes de France (ADF), was set up in 1982 to serve as a clearinghouse for issuing new advances to and receiving repayment of former advances from the public companies. ADF allowed the Government to permit cross-subsidies among companies (as well as within companies as has occurred since the mid-1970s). In 1987 the Government announced its intention to strengthen ADF with an infusion in capital of about €300 million, which ADF could use to make advances to public companies and to increase the State's equity amount. This capital infusion, tapping funds generated by the privatization of Government-owned companies, strengthened the central government's control over the public companies. By the late 1980s both local and national governments began to discuss the possibility of new private concessions on a nonrecourse basis, especially in urban areas, and projects were implemented, for instance, in Marseille (Prado-Carénage Tunnel).

1994–2000: Contract Procurement and Consolidation inside the Public Sector, Extension of Tolls in Urban Areas

Since 1995 (Decree 95-81 mentioned above), multiyear contracts for investment have been implemented; these contracts create a balance between investments and toll increases and give certainty to concessionaires for a 5-year range. The semipublic companies have been consolidated into three main groups in order to profit in terms of geographical coherence and financial viability. Parallel to this consolidation, there has been an increase in capital (from about €4 million to €150 million). From the mid-1990s, tolls started to be used in the cities through specific links and with some difficulties due to acceptability issues. The first intraurban motorway was the Prado-Carénage Tunnel in Marseille (1993), which connected the center of the city (Canebière) to the eastern part of the agglomeration through a tunnel that already existed and just had to be reshaped. The tunnel was franchised to a private company, which ran it successfully. In Lyon, the Western Part of Ring Road so called TEO (1997, about 10 km long) was not as successful; the toll motorway was auctioned and franchised to a private consortium, which levied too high a toll; after demonstrations and protestations of the users, the municipality, COURLY, Le Grand Lyon, cancelled the franchise at a great expense and operated the link by itself with a much lower toll; the operator is EPERLY. Another urban toll link was built in 1998 in the Ile-de-France agglomeration (A14, La Défense – Orgeval, about 20 km long), and worked successfully. A toll motorway link (A86, about 20 km long with an innovative low-gage double-deck tunnel), which will become the second ring road around Paris, is presently under construction.

Since 2000: Extension of Privatization and Commercial Management

More recently, the toll motorways policy has known important changes in order to improve efficiency, according to the general European orientations.[2,3] First, from now on, concessions are granted after being made public and are no longer directly backed collaterally by infrastructure already in operation. Second, the accounting regime of the present concessions was placed more in line with the common one (the core question was the depreciation process), and the State gave up the guarantee for liabilities to existing concessions; in compensation, the duration of the concessions was increased; this State aid was agreed on by the European Commission (State Aid No. N540/2000; letter SG(2000)D/107823, October 10, 2000). As a consequence (linear depreciation in a longer period of time), the companies, without any change in their cash flows, produced positive results, paid income tax, and distributed dividends. Third, tolls were made subject to VAT (CJEC case C-276/97, Judgment of September 12, 2000, Commission/France), for when the use of the road depends on the payment of a toll, the amount of which varies inter alia according to the category of the vehicle used and the distance covered, there is, therefore, a direct and necessary link between the service provided and the financial consideration received. Two new concessions were granted to private firms through this procedure. In March 2002, the main public motorway company, ASF, was introduced on Euronext, the State retaining 51% of the shares, the rest being floated in the stock market. In the same orientation, the motorway firms adopted a more aggressive commercial policy, based on tariff differentiation (discount fares, season tickets, subsidies from local authorities for discount rates to the local users, etc.). APRR was introduced on Euronext in November 2004 through the flotation of 30% new shares and SANEF has been listed since March 2005 under similar conditions. So, at the present time, the three main companies, which are the heads of the three groups mentioned above, are listed. For the last two, it is too early to have an opinion about the evolution of share prices. Nevertheless, it may be noted that, for each initial public offering, the upper limit of the price margin was retained and there was a significant oversubscription. As far as ASF is concerned, quotation is now around €39, while the initial one was €24; this means that the quotation increased by more than 50% in 3 years.

In may 2005 the French Government decided the total privatisation of ASF, SANEF and APRR and the process is ongoing.

Now and After?

At the present time, the French motorways system consists of 7,840 km of toll motorways (75% of the total motorway network) run by:

- six largely State-owned companies on the way to be totally privatised;
- four public-sector-owned companies operating tunnels, bridges, and urban ring roads; and
- four private-sector-owned companies.

The length and route run by each firm are given in Table 4. On the side of financial records, total assets amount to €41.3 billion, of which €3.5 billion are covered by own capital.

The Future?

The evolution of the French motorway concession system is fundamentally a pragmatic process with no dogmas; the main concern was to meet the issues when they occurred to provide the country with the infrastructure needed. In this regard, concession companies appeared to be both a way to

Table 4. French Motorway Companies.

Largely State-owned companies	
ASF	2,325 (listed in March 2002)
APRR	1,801 (listed in November 2004)
SANEF	317 (listed in March 2005)
SAPN	366
AREA	381
ESCOTA	460
Public companies operating tunnels, bridges and urban ring roads	
ATMB	107
SFTRF	67
CCI du Havre	7
COURLY–EPERLY	10
Private companies	
Cofiroute	896
SMTPC (Prado-Carénage)	2.5
Viaduc de Millau	2.5 (December 2004)
ALIS	124 (end 2005)

Note: From French Roads Directorate statistics (values are in kilometres).

earmark resources and a way to provide facilities for investment in the long range through commercial companies fostering private management procedures in the industry, all in the framework of cooperative relationships between the concessionaires and the State. Things took a new turn with:

- The European legislation difficulties to reach the right balance between competition on the one hand, partnership and mutualisation of resources on the other hand;
- A meshed mature network where the different stretches interact with one another;
- A privatisation of the former public sector-owned companies.

As far as French policy is concerned, drawing up of a long-term strategy on the field of motorways is ongoing. Some general tendencies can be stated: First, there is a trend from public management, an old historical tradition, toward "commercial" management; second, the main objective of the French technostructure is the achievement of a good infrastructure network without too much funding from the taxpayer.

Some new elements seem to make the future of the French system take shape:

- First, the French change to greater intricacy in the field of private and public relations is shown by two recent acts: the approval of a new legislation (Ordinance No. 2004-559, June 17, 2004, as approved by Law No. 2004-1343, December 9, 2004) allowing the State and the local authorities to realize more diversified private–public partnerships, such as some kind of shadow toll arrangements, and the establishment of a new agency, the Agence de financement des infrastructures de transport de France, created by decree 2004-1317, November 2, 2004) in charge of the partly funded infrastructure investments of the State through dedicated resources coming mainly from the motorways.
- Second, tolling without concession now in operation in Switzerland, under implementation in Germany for HGV, and under consideration in other countries such as the UK, and the urban road pricing scheme now in operation in London may give rise to prospective new systems of tolling, for instance, zonal pricing or network pricing instead of link pricing, for unbundling concession and toll, and for a service approach of infrastructure provision (Law No. 2004-809, August 13, 2004, Article 20).
- Third, the move from an era of intense investment to the completion of projects and from links to a meshed network is a new challenge. Another challenge is the need for maintenance and more effective and user-friendly operation.
- Last but not least, to set up new relationships between the State and more private and autonomous partners needs a life-long fine-tuning of contracts (concession contract and "contrats d'entreprise") and new skills for a new regulation. From this point of view, the present contracts have a long life: for an overwhelming number of them, their remaining life is more than 20 years long.

In connection with this last point, it is worth noting that finding a right balance between competition and partnership is a new task for both private and public partners and the European commission as well. An approach of regulation based on a negative image of the relationship with private enterprise and, to a certain extent, a mistrust, which implies drawing up contracts as precisely and as short term as possible, is facing practical problems such as the cost of monitoring, the training of regulators, and the setting up of benchmarks. In its aim for the best efficiency and in not respecting dogmas, the pragmatic approach of the French has been largely different, both in its structure and in its set of guidelines. It is founded not on a theory of countervailing powers but on a partnership with contractors vested with a long-term contract (which allows investment with postponed profitability). The use of pragmatic approaches and the trust between the partners are the two main factors.

According to these approaches, France struggles for a specific regulation for toll concessions on the occasion of "Eurovignette" reform negotiations. As a matter of fact, the European Commission's proposal[4] definitively seemed inconsistent with the displayed policies on transport policy and PPP even more so as the control system envisaged aroused a major legal uncertainty. As the Commission's proposal made no difference between a distance-related charge collected on behalf of a public authority (e.g., "Toll collect" in Germany) and a financing toll levied by a concessionaire (e.g., concessions in France, Italy, Portugal, Spain, etc.), the two meanings being bundled into the same term "tolling arrangement,"the same rules were foreseen for determining the levels of tolls to be charged. Member States were constrained to take account of the various costs to be covered, according to a common methodology, and to communicate to the Commission for approval of the unit values and other parameters used in calculating the various cost elements. This system could not fit in with the concession system where tolls are fixed through the markets and contracts and not by a bureaucratic decision. That is why the French government was against this proposal up to the point when concession tolls were submitted to a specific regime in the common position adopted by the EU Council on September 6, 2005.

Above all, concession tolls are a means of financing. The level of tolls is not dramatically different according to the sections on road and the concessionaires; nevertheless, the former stretches are usually charged at a lower toll (mainly because price increase is sensitive) and the average level of toll is higher for a private-owned company (due to a stronger bargaining power). There are some experiences of using toll for traffic management (e.g., motorway A1, north of Paris on a Sunday afternoon), but the revenues for the concessionaire remain stable. As tolling in urban areas is implemented in very few places (e.g., A14, west of Paris, Prado-Carenage Tunnel in Marseille), congestion toll is not an issue at stake in France for the time being.

A double challenge has to be faced:

- To reconcile the effectiveness of provision of public utilities and to meet the specific features of infrastructure (such as risk management and externalities) with the effectiveness of contractual arrangements concluded after bidding (even if the standard competitive market makes it less disputable inter alia because of the lack of actors able to enter), a global comprehensive view from construction of infrastructure itself to services

to final users through maintenance and operation within a long-term contract leaving a reasonable level of freedom to the contractor is needed.
- To "contain" transaction and monitoring costs and to "internalize" regulation.

It can be met by a double transition toward a renewed public–private partnership:

- A transition from standard liberal institutional schemes to a less dogmatic and more cooperative one (with a political but independent regulator) and
- A need for a transition from economies totally led by the State or with a strong State intervention culture (as in France) to more market-oriented practices, with more explicit contractual arrangements.[5]

These migrations seem to be on the way. In its 1994 World Development Report entitled "Infrastructure for Development," the World Bank stressed "using markets in infrastructure;" the 1997 World Development Report is about the "State in a changing world." Coming after a substantial number of reports and communications, the white paper "European Transport Policy for 2010: Time to Decide" (COM/2001/370)[6] advocated public–private partnership and the pooling of revenue from infrastructure charges to solve the headache of funding. But to refer to Max Weber, the major problem with capitalism is not the source of capital but the development of the capitalist mentality, both in private and in public sectors, without forgetting the European Commission, which is the guardian of the market economy.

NOTES

1. Fayard, A. (1980). *Les autoroutes et leur financement.* Paris: La Documentation Française.
2. European Commission. (2000). Interpretative communication on concessions under community law of April 12, 2000. *Official Journal*, C 121, April 29, 2000, 0002–0013.
3. European Commission. (2004) *Green paper on public–private partnerships and community law on public contracts and concessions (COM/2004/0327 final).* Brussels: European Commission.
4. European Commission, Proposal for a directive of the European Parliament and of the Council amending Directive 1999/62/EC on the charging of heavy goods vehicles for the use of certain infrastructures, COM(2003) 448 final, July 23, 2003.
5. *A toolkit for public–private partnership in highways.* Washington, DC: World Bank, 2002 (http://rru.worldbank.org/Toolkits/PartnershipHighways).
6. European Commission. (2001). *White paper European transport policy for 2010: Time to decide.* Brussels: European Commission.

PUBLIC PRIVATE PARTNERSHIPS IN THE IRISH ROADS SECTOR: AN ECONOMIC ANALYSIS

Eoin Reeves

INTRODUCTION

Rapid economic growth in recent years has imposed enormous pressures on the capacity of the Irish economy and brought the critical requirement for upgrading Ireland's physical infrastructure to the fore. As a response, the National Development Plan 2000–2006 (1999), which was published in November 1999, planned for an overall investment of €52 billion in the Irish economy with investment of €22.4 billion specifically earmarked for economic and social infrastructure. One of the prominent features of Ireland's NDP is the promotion of public private partnerships (PPPs). In the run-up to the publication of the NDP, the Irish government, in conjunction with business groups, actively promoted the use of PPPs as a vehicle for providing large-scale infrastructural investment. As a consequence, €2.34 billion of investment using the NDP was earmarked for procurement using the PPP model.

The biggest element of both overall and PPP-specific investment within the economic and social infrastructure operational programme (ESIOP) of the NDP is in the roads sector. This chapter details the planned contribution of the PPP model to Ireland's road investment programme under the NDP.

Procurement and Financing of Motorways in Europe
Research in Transportation Economics, Volume 15, 107–120

ISSN: 0739-8859/doi:10.1016/S0739-8859(05)15009-4

More specifically, it examines the experience with the PPP model in the roads sector in terms of the economic theories of relevance to the PPP model of procurement and explicit government objectives.

BACKGROUND: IRELAND'S INFRASTRUCTURE DEFICIT AND THE NATIONAL DEVELOPMENT PLAN

Ireland's low stock of quality infrastructure is well recognised. A period of fiscal stabilisation in the 1980s meant a necessary curtailment of the capital expenditure programme and expenditure (in real terms) on Ireland's public capital programme (PCP) fell each year over the period 1982–1989. Although the rapid growth of the Irish economy since the early 1990s resulted in a convergence towards EU living standards (measured by GDP per capita), accumulated wealth in terms of physical infrastructure and accumulated human capital remains considerably lower than that for countries at or above the EU average income levels. Prior to publication of the NDP the Department of Finance (2000) published details of the country's infrastructural deficiencies in sectors such as roads, railways and environmental facilities. With regard to roads, the report highlighted the poor quality of the motorway network. It stated that by 1996 the network had reached 13% of an EU index, weighted for population and land area, which is by far the lowest figure for any EU member state. Within this context, the NDP originally set out plans for overall investment of €22.4 billion in economic and social infrastructure with investment in national roads sector accounting for 27% of overall planned expenditure.

IRELAND'S ROAD INVESTMENT PROGRAMME

In their evaluation of investment in the Irish road network, Fitzpatrick and Associates (2002) describe the network in terms of two road classes – 'national' and 'non-national' roads. National roads are divided into the subcategories 'national primary' routes (i.e. major long-distance through-routes linking the principal ports and airports, cities and large towns) and 'national secondary' routes (i.e. medium-distance through-routes connecting important towns, serving medium to large geographical areas and providing links to national primary routes). Non-national roads are all other roads in the network, consisting of regional roads (i.e. feeder routes into, and providing main links between, national roads) and local roads (including such roads in urban areas).

Prior to the publication of the NDP, national roads accounted for 6% of the total road network and 38% of all road traffic, with non-national roads accounting for the remaining 94% of road network and 62% of all road traffic. In relative terms, Ireland's road network is very distinct by international standards. Prior to the publication of the NDP the road network was extensive relative to the population. However, just 0.3% of the network was of motorway standard compared to the EU average of 1.17% (Fitzpatrick & Associates, 2002). The *National Roads Programme 2000–2006* in the NDP, therefore, seeks to increase the proportion of the network that is of motorway standard and can be viewed as the current roads programme. The contents of the programme is divided into five groups of schemes (or categories of projects) according to the class of route or nature of the projects involved: (1) major inter-urban routes (MIUs); (2) Dublin area projects; (3) other national primary routes; (4) national secondary routes and (5) support measures. It is worth noting, however, that most of this programme is specified in terms of routes requiring improvement rather than specific projects. Projects have only been specified and time-bound for completion by 2006 in the case of MIUs (31 projects) and Dublin area projects (five projects).

PUBLIC PRIVATE PARTNERSHIPS AND THE NATIONAL ROADS PROGRAMME

The identification of the PPP model of procurement for a number of individual road projects is a key feature of the *National Roads Programme 2000–2006*. In June 1999, the government identified three road projects (with the added commitment of ensuring the potential of a fourth named project) for the purpose of piloting the PPP model. The PPP programme was subsequently extended and currently consists of ten PPP contracts. In 2000, it was estimated that they represented investment of €1.3 billion (2000 prices), which included a potential private finance investment of €889 million (Fitzpatrick & Associates, 2002, p. B63).

All PPP road projects are planned as concession contracts. The private consortia are responsible for the design, build, operation and finance elements of the projects. Payment will be secured via a combination of direct tolls and unitary payments by the public sector. The National Roads Authority (NRA) is the primary authority for the day-to-day delivery of the NDP roads strategy on behalf of the government. It conducts the procurement process in the case of all PPP projects and it acts as the contracting party in contrast to conventional contracts, which are signed by the local authorities. According to Fitzpatrick and Associates (2002, p. B62), the NRA established the following key principles for guiding its policy on PPP schemes:

- Only schemes that were not already so far advanced under traditional planning and procurement process that the PPP procurement would significantly delay their delivery were considered for selection.
- The alternative toll-free route (to the PPP tolled scheme) had to be available for users.
- Tolled roads had to be spread across the main national routes in order to create an equitable distribution of user charging on the newly constructed network.
- A project would have to be of sufficient size to project value for money (VfM) in the PPP process (a cut-off of €38 million was used).
- The aim was to secure complete projects from the private sector but where necessary, a public subsidy will be considered for high cost schemes that cannot be solely financed from hard tolls.
- Only hard tolling, where the road-user pays directly, will be considered.

Given the unique features of the PPP model of procurement such as the long duration of concession contracts (30 years) and hard tolling arrangements, this chapter examines the experience to date with road PPPs. In particular, this chapter analyses the experience in terms of economic theories of relevance to the PPP model of procurement.

ANALYSING PUBLIC PRIVATE PARTNERSHIPS

Over the last 20–30 years, governments worldwide have sought new means of improving the delivery of public services. A common characteristic of many of the arrangements that have been adopted has been the engagement of the private sector in the delivery and/or financing of public services and this has led many governments to experiment with reforms such as contracting-out and PPPs.

There is no single definition of PPPs but for the purpose of this chapter a PPP is described as an agreement between the public sector and a private sector company to provide an asset or service, which would traditionally be provided by the public service, but as part of a PPP project will be provided by the private sector or jointly by the public and private sectors. The essence of a PPP project is that the private sector will do one or more of the following: (1) provide private finance to fund the project; (2) enter into a long-term service contract; (3) undertake the design and construction of an asset on the basis of an output specification prepared by the public sector and designed to meet broad performance targets; (4) enter into a joint venture arrangement with the public sector to provide a service or asset. This differs from the traditional model of procurement practice in the public sector where the construction of publicly owned assets such as roads or prisons is typically carried out to detailed specifications by private contractors following a competitive tender. The principal PPP models are as follows: (1) Design and build (DB) – where the private sector is contracted to design and construct a facility which, upon completion, is transferred to public ownership. The private sector contractor(s) receive payment from the public sector. (2) Design, build and operate (DBO) – where the private sector is contracted to design, build and operate an asset. The project is fully financed by the public sector. (3) Design, build and operate and finance (DBOF) – where the private sector funds the capital investment and recovers the costs over the life of the project (usually over 30 years) via payments from the public sector, e.g. 'shadow toll' road schemes. (4) Concession – where payment or part payment for the service is provided through direct charging of the end user, e.g. hard toll road schemes.

CONTEXT FOR EXAMINATION – ECONOMICS OF CONTRACTING AND PPPS

Sappington and Stiglitz (1987) devised a 'privatization theorem', which provides a useful basis for examining the economics of PPPs. These writers set out conditions under which a model of procurement such as PPP allows government to achieve three principal objectives, i.e. efficiency, equity and rent extraction (i.e. elimination of excessive private profits). In essence, these conditions involve an auction system that requires bidding from two or more risk-neutral forms that have symmetric beliefs about the least-cost production technology. Despite the simplicity of this system the authors assert that there are a number of reasons why the ideal outcome will not be generally attainable in practice. These can be reduced to (1) difficulties in extracting rents from the chosen producer and (2) the cost of negotiating, monitoring and enforcing contracts (i.e. transaction or agency costs). While the Sappington and Stiglitz theorem was not devised in the specific context of PPPs it provides a useful basis for deriving criteria for examining this form of contracting. These criteria include: (1) competition for the market; (2) efficiency/value for money; (3) risk allocation; (4) innovation and (5) rent extraction.

Economic Criteria for Examining PPPs

Competition for the Market

For many years, economists have articulated the case for competition in the context of tendering and franchising. The benefits of competitive tendering in terms of improvements in productive efficiency (cost reduction) are well documented (for a review, see Domberger & Jensen, 1997) and there is the expectation that more intense competition increases efficiencies. Conversely, contracting problems can arise as a result of a limited pool of bidders. Williamson (1975) explores the costs of 'small numbers exchange'. In these imperfect markets, it may be difficult to drop those who have behaved opportunistically in the past at the time of renewing contracts.

Efficiency and Value for Money

For policy makers, the potential for accruing efficiency gains is a major attraction of the PPP model. Governments commonly justify the adoption of the PPP model on the grounds that it achieves better VfM compared to conventional procurement. In Ireland, the *Framework for PPPs* (2001) – the definitive government statement of the scope, goals and principles of the PPP programme – stresses the objective of "value for money for the tax-payer, inter alia, through optimal risk transfer and risk management" (2001, p. 3).

Risk Allocation

Risk allocation is an important issue in the economics of contracting. The economic theory of principal and agent focuses on the design of 'optimal contracts' in the face of differences (asymmetries) in the information and objectives of contracting parties. Emphasis is placed on the optimal allocation of risk as a means of 'incentivising' agents to achieve principal's objectives. The case for PPPs in Ireland is explicitly articulated in terms of transferring appropriate risk levels to the private sector. In PPP contracts, this generally takes the form of identifying categories of risk and agreeing on whether they are borne by the client (principal) or contractor (agent). Moreover, as PPP contracts can link different elements of infrastructure projects (e.g. link the design and construction with one or all of the finance, operation and maintenance elements) there is better scope for transferring risk compared to traditional procurement methods.

Innovation

As the PPP model involves the delivery of public services on the basis of an output specification prepared by the public sector and designed to meet broad performance targets it is argued that PPPs encourage private sector innovation and improved service quality. Under traditional procurement the public sector specifies the asset to be built (e.g. school building). Once this is completed the public sector assumes responsibility for its continuing operation and maintenance. Under PPP, the public sector provides an output specification wherein the requirements for the service to be provided are specified. This allows competing bidders the scope to create innovative solutions that may offer better VfM.

Rent Extraction and Re-financing

Sappington and Stiglitz (1987) specify rent extraction as a government objective in the context of privatisation. Under certain conditions, the scope for excess private sector profits is minimised. Evidence from PPPs in the UK, however, highlights the problem of imperfect rent extraction due to aspects of deals to re-finance PPPs. When the construction stage of some projects has been completed some contractors have returned to capital markets and re-financed deals at significantly lower costs. The resultant financial gains represent economic rents.

ANALYSING IRELAND'S ROADS PPPS

Speed of Delivery

In the Irish context, one of the principal reasons for adopting the PPP model has been its potential for speeding up the delivery of infrastructural assets and related services. Advocates of the PPP model argue that investment in infrastructure under PPP is more time and cost-efficient than under traditional procurement. This is largely due to the integrated nature of PPP models, such as DBOF, and 'concession' contracts. It is argued that the 'bundling' of design with project execution and maintenance helps align the incentives of parties responsible at different stages of the investment cycle, which can provide a basis for more efficient and timely service delivery. Furthermore, the PPP model provides a basis for designing incentives within the terms of the contract between client and contractor (e.g. make payment conditional on completing certain stages). Against this, however, it must be noted that the PPP approach involves a negotiating process, which is longer, more complex and more expensive than conventional projects.

In the case of Ireland, movement through the stages of procurement of road PPPs has been slower than expected and the PPP model has failed to speed up delivery of roads. Of the 11 projects that were originally nominated as PPPs, one has been re-designated for conventional procurement. Another project was effectively an extension of a concession that pre-dated the official launch of the PPP programme while two PPP projects remain at the pre-procurement stage. Hence, the analysis in this chapter is confined to seven projects that have progressed to different stages of the procurement process. Table 1 provides details with regard to the three projects where contracts have been signed and construction has commenced while Table 2 details the four projects that are at the early stages of procurement (announcement of pre-qualification tenders).

Overall, the level of progress to date indicates that the targeted date for completion of the PPP programme is 2009 – 3 years behind schedule. This delay can be attributed to a number of reasons, some of which are not necessarily attributable to the characteristics of the PPP model. These include factors that have contributed to a significant slowdown in delivering the overall roads programme under the NDP such as the extraordinary rise in the cost of road building since the NDP was launched. The NRA has estimated that the cost of building a kilometre of road has doubled since 1999 and the entire cost of the road-building programme is now forecasted at €15.7 billion compared to an initial forecast of just under €6 billion. This is attributed to soaring land acquisition costs and inflation in the building sector, which rose by 15, 10 and 5% in the years 1999, 2000 and 2001,

Table 1. PPP Projects Under Construction.

Route	Length (km)	Date of Award	Projected Completion Date
N4 Kilcock-Kinnegad	11	March 2003	Autumn 2006
N1/M1 Dundalk Western Bypass	11	February 2004	Spring 2006
N8 Rathcormac Fermoy Bypass	18	June 2004	Summer 2007

Table 2. PPP Projects Under Procurement.

Route	Length (km)	Current Stage	Next Event
N25 Waterford Bypass	37	Awaiting BAFO tenders	Contract award
N3 Clonee Kells	75	Awaiting ITN submissions	Shortlist ITN tenders
N7 Limerick Southern Ring Phase II	10	Announcement of pre-qualification tenders	Commencement of tender process
M50 Upgrade	24	Announcement of pre-qualification tenders	Issue of ITN documentation

Notes: BAFO, best and final offer; ITN, invitation to negotiate.

respectively. Other factors, such as measures taken to prevent the spread of Foot and Mouth disease in 2001, also caused slippage by delaying site investigations and the Environmental Impact Statement (EIS). With specific regard to the PPP model, one of the principal reasons for delays in completing projects is the length of the procurement process. In an early examination of PPP projects in the roads sector, Fitzpatrick and Associates (2002) estimated that the PPP procurement process takes around 5 months longer than traditional contracts due to the requirement for additional stages to the process such as the "Public Sector Comparator" stage.

The NRA asserts that measures are being taken in order to speed up project delivery under all methods of procurement. An important example is the decision by government in the 2003 budget to implement a multi-annual funding arrangement to replace the long-standing practice of an annual budgetary allocation process. This new arrangement commits to continued substantial investment in national roads with exchequer funding of €7 billion guaranteed over the years 2004–2008. In addition, Murphy (2004) points to new legislation that has been passed to resolve problems regarding archaeological national monuments and measures have also been taken to streamline the statutory approval process (e.g. consideration of EISs).

Value for Money

Competition for Contracts

On the basis of number of pre-qualification submissions the early indications were that the market for contracts in road PPPs were competitive with over ten expressions of interest for the first three projects. The norm has been to invite three to four submissions, thereby encouraging a competitive procurement process. In each of the contracts signed to date two detailed bids were received.

It is noteworthy, however, that the number of expressions of interest has fallen as deals come to the market. Whereas there were 12 pre-qualification submissions in the case of the Waterford Bypass project numbers have fallen to seven and five in the cases of the N1/M1 and N3 routes, respectively. Moreover, only two expressions were received for the N7 route although it should be recognised that this project includes tunnelling under the River Shannon and therefore requires a comparatively specific investment. Nevertheless, the observed decline in pre-qualification submissions has raised questions over the sustainability of competition for PPP contracts. In one review of private sector experiences with PPP competitions, O'Rourke (2003) found that high levels of tendering costs and risk transfer demands were two important factors in this respect.

Risk Transfer

In relation to risk transfer, one of the biggest elements of risk concerns the model of tolls. Compared to countries such as Spain and Portugal, who led the field in PPP toll road projects in Europe, a striking feature of Ireland's programme is the use of hard tolls as opposed to shadow tolls. Besides the risk of alienating the road-using public, hard tolls increase the risk borne by the concessionaire who must take usage risk (Project Finance, April 2001). Beyond this fundamental concern a number of aspects of the Irish PPP roads programme have been the subjects of criticism from the private sector. According to O'Sullivan (2003, p. 41) these include:

- retention by government of the power to set and reset tolls;
- reservation by government of the right to order the concessionaire to upgrade to variant tolling mechanisms;
- absence of compensation to provide for termination due to operator default;
- transfer of significant competing route risk as the state has maintained the commitment to provide a toll-free alternative route between all destinations (although this is met already by existing roads around the concession areas).

The validity of the first two criticisms has been undermined by the fact that contracts are now signed for four PPP contracts suggesting that the due diligence process undertaken by funding agenies have indicated that sufficient protections were in place. On the question of compensation for termination, it should be noted that this clause is standard on road deals under the Private Finance Initiative in the UK. With regard to the issue of alternative routes it should be stressed that while the key risk is that local authorities are free to duplicate infrastructure, the practical reality is that they are poorly resourced and therefore unlikely to present concessionaires with such risks.

Private sector concerns over the question of risk transfer are an inevitable feature of the PPP experience. The early indications are that the NRA has pursued an exacting policy on risk sharing with some of the transferred risks particular to the Irish case (e.g. alternative route risk). The argument that this approach to risk transfer threatens the competition that is fundamental to the PPP model appears less plausible as more contracts are signed and as the extent of debt syndication increases, which has been the case to date in Ireland.

Innovation

As the assets and services provided under the PPP model are designed on the basis of output specification, as opposed to prescriptive input specification, there is potential scope for greater private sector innovation. One review of the Irish roads PPP programme to date conducted by O'Rourke (2003) found that it is the unanimous view of all those interviewed that the amount of design input from the private sector is minimal. Despite their description as "Design, Build Finance and Operate" projects the current crop of road projects remain largely building projects…the emphasis in the roads programme, according to contractors, remains resolutely on input specification as in traditional procurements (2003, p. 9).

Two of the principal obstacles commonly identified are the detailed and prescriptive nature of the EIS and a lack of flexibility in the planning process in Ireland. It should be noted, however, that these restrictions apply to all the NRA's procurement processes and are not solely related to the PPP model. Nevertheless, if scope is to be created for innovation, which is crucial to the case for PPPs, significant reforms of the statutory approval process are required.

Evidence of Value for Money

The objective of VfM has been central to the case for PPPs in Ireland. Much of the commentary on PPPs focuses on the measurement of VfM in the strict financial sense. This measurement, however, is an exercise that is highly subjective and sensitive to assumptions with respect to cost and value. In Ireland, VfM is tested by comparing the net present value of the PPP project with the so-called public sector benchmark (PSB). The PSB represents the hypothetical cost of providing the facility using conventional means of finance. This comparison provides an ex ante estimate of VfM and whether this is accrued over the life of the contract is not known.

If VfM is to be achieved the PPP model must yield efficiencies that outweigh the higher costs of borrowing faced by the private sector as well the costs of the tendering process. Detailed information in relation to calculation of the PSB is not published in Ireland as a matter of government policy. In the case of roads, however, the NRA provides the bottom-line PSB figure, which is compared with the financial details of the agreed PPP contract. The details in Table 3 suggest that the PPP model yields sizeable cost-savings for the Irish exchequer. Any assessment of VfM must, however, include the toll revenues foregone by the state. As this will depend critically on factors such as traffic flow and subsequently the level of toll revenues; the question of VfM remains uncertain at the early stages of PPP contracts.

Rent Extraction/Re-Financing

The need to appropriate a share of the gains from re-financing PFI/PPP deals has been one of the principal lessons learned by policy makers in the UK. In 2002, the National Audit Office reported that 91% of contracts signed since 2001 include mechanisms for sharing the gains from re-financing PPP deals. The comparative figure for contracts signed before 2001 was 54%.

In interviews conducted for the purpose of this research, NRA officials explicitly referred to taking the UK experience into account when negotiating PPP contracts. As a result, the NRA has successfully extracted rents by negotiating the sharing of gains from refinancing PPP deals. In each of the three contracts signed to date the sharing of gains on a 50:50 basis has been agreed.

CONCLUSIONS

The Irish government has shown considerable commitment to the PPP model of procurement as a means of addressing the country's deficit of physical infrastructure. The experience to date has not, however, been positive. Six years after the announcement of the first PPP projects only a small number of contracts have been signed (in the roads and schools sectors) and

Table 3. Details of Payments on Irish PPP Contracts.

Project	PSB (€ million)	Payments to Concessionaire (€ million)	Payment Details
N4 Kilcock-Kinnegad	550	152[a]	€146 million over period of construction €6 million during period of operations Share of tolls paid to the State
N1/M1 Dundalk Western Bypass[b]	340	0	All tolls on Dundalk Bypass accrued by operator 95% of tolls on existing motorway paid to the State during construction (valued as €18 million). A share of future revenues thereafter paid to the State
N8 Rathcormac-Fermoy Bypass	320	120	€80 million over period of construction €40 million during period of operations Revenue sharing applicable after traffic volumes exceed 21,000 vehicles per day

Source: NRA Project Tracker, http://WWW.nra.ie/PublicPrivatePartnership/ProjectTracker.
Note: All values in 2003 prices.
[a]Excludes land/preparatory costs.
[b]This contract involves operation and maintenance of an existing motorway and procurement of Dundalk bypass as concession PPP.

the government's implementation of the PPP programme has been the subject of fierce criticism from the private sector and other sources. These difficulties reached a critical point in November 2004 when the Minister for Finance excluded capital spending from the estimates of government spending for 2005 citing the slow roll-out of the PPP programme as the reason for postponing the announcement of relevant details.

The roads sector accounts for most PPP activity to date (measured in terms of investment expenditure). The details presented in this chapter reveal that the PPP programme is at least 3 years behind schedule. This has been attributable to a number of factors, not all of which are specific to the PPP model of procurement. Whether PPPs achieve desired outcomes such as faster delivery of infrastructure and related services while achieving VfM remains to be seen. The experience to date indicates that the degree of competition for contracts is being threatened by loss of private market enthusiasm due to factors such as high bidding costs. Moreover, there is little to suggest that the model is providing scope for private sector innovation, which is central to its legitimacy. On a positive note, the indications are that the NRA is achieving significant risk transfer and that the contracts, which have been signed, are 'bankable' with providers of finance achieving syndication of debt. Whether VfM is achieved in the long run ultimately depends on traffic flows and the precise details of revenue sharing.

REFERENCES

Department of Finance. (2000). *Ireland, community support framework, 2000–2006*. Dublin: Stationery Office.

Domberger, S., & Jensen, P. (1997). Contracting out in the public sector: theory, evidence and prospects. *Oxford Review of Economic Policy, 13*(4), 67–79.

Framework for Public Private Partnership (PPPs). (2001). *Framework for public private partnership*. Dublin: Stationery Office.

Fitzpatrick & Associates. (2002). *Evaluation of investment in the roads network*. Dublin: Stationery Office.

Murphy, G. (2004). Road testing. *PFI Intelligence Bulletin* (September), 10–11.

National Development Plan 2000–2006. (1999). Dublin: Stationery Office.

Sappington, D. E. M., & Stiglitz, J. E. (1987). Privatisation, information and incentives. *Journal of Policy Analysis and Management, 6*(4), 567–582.

O'Rourke, C. (2003). *Public private partnerships in Ireland – How they can be streamlined*. Dublin: Construction Industry Federation.

O'Sullivan, D. (2003). *Cintra's Irish rebel song*. Project Finance International.

Williamson, O. E. (1975). *Markets and hierarchies: Analysis and antritrust implications*. New York: The Free Press.

HISTORY AND REGULATION OF ITALIAN HIGHWAYS CONCESSIONAIRES

Andrea Greco and Giorgio Ragazzi

DEVELOPMENT OF THE ITALIAN HIGHWAYS NETWORK: A SHORT HISTORY

The word 'autostrada' was used for the first time in 1922 by Mr. Puricelli, an entrepreneur who put forward the idea of building new roads for exclusive use of motor vehicles (the first such project was that for 'Autostrada dei Laghi', north of Milan).

In less than a decade, starting in 1925, 375 km of highways were built by several private companies, under concession by the state. But traffic was not enough to cover costs: in 1930 there were only 250,000 motor vehicles in the whole country! To save licensees from financial collapse, concessions were taken over by a government agency, investments were stopped and tolls reduced by one-third. In 1940, the highways network amounted to 485 km, of which 174 under concession to private companies and the rest managed by a state agency.

In 1950, the government-owned holding group IRI was commissioned to carry on the project for the highway link between Milan and Naples, through its subsidiary 'Autostrade Concessioni e Costruzioni'. Eight years later the new highway was opened.

Procurement and Financing of Motorways in Europe
Research in Transportation Economics, Volume 15, 121–133

ISSN: 0739-8859/doi:10.1016/S0739-8859(05)15010-0

Highways were financed by companies owned by the government or by local public institutions, under concession contracts granted by ANAS, the government agency responsible for state roads. A law (463/1955) set the principle that every highway should be self-financing, with government contributions limited to a maximum of between 20 and 30% of the total cost (36% for the Milan–Naples highway). It was also stated that if toll revenues exceeded the forecast of the initial financial plan, the concessionaire was to devolve the excess to the state (keeping only 10% of the additional income).

In 1961, a new law was approved introducing a regulation based on the cost of service and further reducing the independence of the concessionaires. The transport minister was empowered to set the level of tolls; the government offered to guarantee licensees' debt up to 50% of investment costs (later increased to 100%) and to increase its subsidy up to 52% of the cost of the new highways. However, concessionaires were to hand over to the state toll revenues in excess of agreed costs and of their own capital remuneration, fixed at 6.5%. The state, through IRI, assumed a more prominent role: most new investments were assigned to IRI's Autostrade, which was however to turn over to the state the whole network in 2003, according to a convention signed in 1968.

By 1970, Italy had a very good network of highways (3,913 km), more than twice that of France and three times that of the UK. In 1975, 5,000 km of highways were completed, 665 km were under construction and 1,024 km were planned; 52% of the network was operated by Autostrade, 42% by companies controlled by local public institutions and only 6% by private companies.

Concessionaires' own capital covered only a minimal part of investment costs: thanks to the state guarantee they had easy access to credit, both in the bond market and from banks. This changed drastically in the mid-1970s due to the petrol crisis and the increase of interest rates and investment costs. Concessionaires' finances came under strain, at a time when the state was also facing serious financial imbalances. Therefore, in 1975 the government decided to stop construction of new highways (law 492/1975), although projects underway were to be completed. By 1980, the network had increased to 5,900 km. Since then, very little has been added to the network's length: 20 years later the total length was 6,478 km.

Tariffs were often frozen by governments, as part of anti-inflationary policies. This was possible because almost all licensees were owned either by IRI or by local public institutions. However, freezing tariffs reflected in a worsening of the public sector indebtedness, as licensees' debts were guaranteed by the state. In substance, this sector was regarded as part of the public sector, until the end of the 1990s, when Autostrade was privatized and new conventions were drafted with the other concessionaires.

THE PRIVATIZATION OF AUTOSTRADE SPA

The company Autostrade built over one half of the Italian network, in little over two decades, up to the end of the 1970s. This was all financed through debt: IRI's original capital contribution to Autostrade was negligible. The company, originally conceived as an instrument for building highways on behalf of the state, had become highly profitable in the 1990s and distributed fat dividends to IRI. Autostrade's concession was due to terminate in 2003, and most of its investments had been amortized by the end of the 1990s.

IRI (and the state itself) was, however, in a shaky financial situation, and the government decided to launch a vast programme of privatizations that would have ultimately resulted also in the dismantling of IRI (Baldassarri, Macchiati, & Piacentino, 1997). As a part of this programme, Autostrade was privatized at the end of 1999. To make this possible, and to maximize the selling price,[1] a new convention was drafted extending Autostrade's concession to the year 2038 (a very rich gift indeed!) and including very generous provisions for future tariff increases. Pressures from other concessionaires forced the government to grant them too extensions of their concessions: indeed, all concessions were extended at the end of the 1990s.

According to the 'project financing' philosophy of the laws that had regulated construction of highways, in 2003 tolls should have been abolished, or reduced to cover only operating costs. The decision to maintain instead the same level of tolls (with subsequent adjustments for inflation) for an additional long period of time amounted in fact to introducing a new tax on highways users, which was reflected in a assured flow of profits for the company. The price paid by private investors was justified as the present value of this expected flow of profits: in other words, one could say that investors purchased from the state the right to collect this new tax, for 40 years!

IRI cashed 6.6 billion euros from the sale of its 87% of Autostrade. A controlling stake of 30% was purchased by Schemaventotto, a company controlled by the Benetton family. In the following years, Autostrade was the star of the Italian stock exchange; the price of its shares trebled. In 2003, Schemaventotto bought from the market an additional 32% of the capital, at the cost of about 8 billion euros, through Newco which was subsequently funded into Autostrade, which is thus to repay such debt out of toll revenues.

THE PRESENT SITUATION OF THE NETWORK

Italy had as of 2004, 5.593 km of tolled highways, under concession to 24 licensee companies, and 894 km of free highways managed by ANAS.[2] All free highways are in the south of Italy, for political as well as economic reasons.

The Mezzogiorno being a 'poor' region, governments have traditionally tried to promote its development by increasing public spending, especially in transport infrastructures (thus, for instance, Sicily has the longest network in Italy), even if the volume of traffic was relatively low and often insufficient to covers costs through tolls. Some of the free highways, like the never completed Salerno–Reggio Calabria, have a very low standard. However, the issue has now become politically sensitive, with parties based in northern Italy arguing that the same rules should be applied countrywide, with regard to payments for the use of highways.

ANAS is actually planning to introduce tolls on all its highways, as upgrading works are completed, but no clear government decision in this sense as yet been taken, perhaps due to its potential political repercussions. A first example in this direction was the 'privatization' of the Autostrada dei Parchi, 115 km of roadway connecting Rome to Pescara and Teramo. The quality of this tract was poor, but tolls were low. After an open tender, the concession was assigned to a subsidiary of the Autostrade group, which undertook to upgrade the infrastructure and to pay (over a long period of time) 1.4 billion euros to ANAS; as part of the agreement, tolls were increased by 50%. Users complained because tolls were increased within a short period of time, while upgrading investments are still at the project stage.

There are two main operators: Autostrade, which manages, with its subsidiaries, 3,400 km of toll road (some 60% of the total), and ASTM-SIAS with about 1,000 km of highways plus other 138 km managed by companies where ASTM has an important share of capital. Both groups are privately owned: the Benetton family controls the first one and Mr. Gavio controls the second one. While Autostrade has one company ('Autostrade per l'Italia') which controls most of the network (2,854 out of 3,400 km of the whole group), ASTM-SIAS controls six subsidiaries, none of which has tracts much longer that the others.

The third operator is Autostrada del Brennero, which is owned by local public authorities and manages 314 km between Modena and the Austrian border.[3]

There are other six independent companies, owned by local public authorities or by public–private partnerships.

There are no technical or economic reasons to justify the concentration of half of the network under management by a single operator, Autostrade. However, fragmentation is very relevant for the way in which new investments are financed. The 1997 contract between ANAS and Autostrade, which included investments for 4,235 million euros for the period 1997–2003, did not foresee any government subsidy. Moderate increases in tolls spread over its entire network were sufficient to finance the new investments (most of which are still now to be started). Different was the case for an Autostrade's subsidiary that manages the Torino–Savona tract. This highway was the last one to have, for some tracts, a single lane. In 1999, it started to double the carriage and to improve service and safety; its low volume of traffic was however not sufficient to cover investment costs, and thus it obtained a government subsidy to cover about one half of the cost. Although Autostrade owned 99.9% of this company's capital, it did not cover the financial cost of this project, because it was formally carried on by a different company. Fragmentation is thus a system through which groups can maximize government subsidies.

The ASTM-SIAS group has also obtained large subsidies, amounting to over 1 billion euros, to cover the investment programs of some of its subsidiaries. Very substantial toll increases were in the meanwhile granted to this group, again to finance new investments, but the planned investments have so far not yet materialized, and it is not clear how the resulting extra profits will be accounted for by ANAS, over the convention period.

Tolls average (2003 data) 5.8 euro cents per kilometre, but they vary between 14.0 and 4.6 euro cents, depending on the concessionaire. It is estimated that the cost per kilometre of an average passenger car is roughly equivalent to the cost per kilometre of the gasoline tax: tolls amount to a doubling of the gasoline tax, for highways users.

Total revenues from highways tolls amounted, in 2003, to 4.7 billion euros. After deduction of 20% for VAT and about 3% for a central fund established in the 1970s to cover default risks on concessionaires' debts, net revenues accruing to the concessionaires amounted to 3.8 billion euros. If we consider income taxes also on companies' profits, we may conclude that close to one-third of gross revenues from tolls ends up in the state budget.

Another large share of revenues goes into (gross) profits: 20% in the case of Autostrade, 30% for Autobrennero (including a tax free fund, see note 3 at the end of the chapter), over 40% for Autostrada Torino-Milano, just to mention the major operators. In the period between 1997 and 2003, revenues of Autostrade rose from 1762 to 2571 million euros and profits increased from 151 to 522 million euros, i.e. from less than 10 to 20% of revenues; revenues of Autostrada Torino–Milano rose from 78 to 126 million euros, and profits increased from 18 to 41% of revenues, even if, in both cases, investments during the period were very few and total length remained unchanged.

The continuing growth of traffic has induced the government to approve, in the year 2000, the construction of new highways listed in the General Transport Plan and subsequently confirmed in a law ('Legge Obiettivo').

The plan envisages investments for roads of 48 billion euros over a 12-year period, of which 26 billion is in the Mezzogiorno.[4] However, financing for this ambitious program has still to be arranged. Apart from concessionaires' planned investments, state budget funds are very limited. The government intends to raise as much private capital as possible, including borrowings by a government owned financial company not consolidated in the public sector (its debt would thus be excluded from the definition of 'Maastricht' government debt). Private companies are allowed to propose new transport infrastructures projects, on a self-financing basis.

Important changes in regulation and legal framework have also been introduced. Building new roads (and railways) has always been difficult, due to the need to obtain approval by the majority of the municipalities involved, which often demand costly compensations and require lengthy negotiations. To speed up construction, the veto power of municipalities has been abolished, for a selected list of priority projects. Nonetheless, the objective to introduce a 'fast lane' for priority projects has not been achieved so far, due to long delays in project preparation as well as lack of government funds.

REGULATION

ANAS: Player and Regulator

ANAS is a state-owned company, legally organized as a share company. Its traditional main task it to build state roads, and to maintain some 20,000 km of state roads. In addition, ANAS has been assigned the task of enforcing regulation of highways concessionaires, on the basis of a 'convention' with the Ministry of Transport. It is to control the licensees, negotiate renewal or extensions of concessions, the level of tolls and all the parameters included in the price cap structure. New investments must be approved by ANAS both from the economical and engineering aspects.

However, ANAS itself participates in the capital of several highways concessionaires and of companies promoting the construction of new highways. This potential conflict of interests has come under criticism, and some experts and politicians have proposed to assign regulation to an independent authority (Boitani & Petretto, 1999), but the competent ministry opposes this project.

Renewal and Extension of Concessions

Renewal of concessions is a most delicate aspect of regulation. Concessions have always been renewed over time: some of Italy's 24 concessionaires still operate tracts built as far back as the 1930s.

Most concessions were renewed at the end of the 1990s, together with the privatization of Autostrade. The main reason brought to justify renewals was the fact that, in the past, tariffs had been frozen for various years, and licensees claimed a credit towards the state for revenues thus lost.[5] Renewals were also considered a way to compensate licensees for other credits they claimed to have towards the state or ANAS, and to finance new investments (most of which are still now to be undertaken).

Concessions were renewed on the basis of a financial forecast made to assure and control the profitability and financial balance of each concessionaire, with tariff revisions planned every 5 years. Financial plans are not made public, since conventions are considered contracts between two 'private' companies (ANAS is indeed formally a share company). Therefore, it is very difficult for consumers as well as for members of parliament, to control the appropriateness of extensions and/or tariff levels.

When Autostrade's concession was renewed, the European Commission raised various objections, in particular it asked that the management of the network be separated from construction and that no subsidiaries of Autostrade should participate to the construction auctions.

Another reason for extensions may be to amortize new investments agreed with ANAS. The European Commission in increasingly objecting to such practice, demanding that concessions be assigned through tenders, when they expire. This is now facilitated by a norm ('direttiva Costa-Ciampi', 1998) that requires the winner to pay to the incumbent the cost of investments not yet amortized, although no such case has yet been experienced.

The Price Cap

Up to 1996, the principle followed for yearly tariff adjustments was as follows:

$$\Delta T = \Delta P - \Delta V(1 - \alpha) \tag{1}$$

where ΔT in the increase of tariff, ΔP stands for inflation, ΔV is the increase in the volume of traffic and α is the percentage increase of operating costs due to an increase of traffic. The licensee had no 'traffic risk', i.e. any increase of traffic was to be reflected in lower tariffs. The tariff level was to be set so as to assure an adequate profitability, calculated on a weighted average cost of capital (WACC) to be determined yearly by the government, based on prevailing financial market conditions.

In view of the privatization programme, Cipe (Interministerial Committee for Economic and Financial Planning) decided, in 1996, to adopt price cap as a general criterion to adjust tariffs for the public utilities, including highways (Iozzi, 2002).

In this sector the increase in toll is to be regulated as a function of three factors:

$$\Delta T \leq \Delta P - \Delta X + \beta \Delta Q \tag{2}$$

where ΔT is the increase of tariff (weighted average for the entire network of each concessionaire), ΔP stands for (planned) inflation,[6] ΔX is the planned increase in productivity, ΔQ is the percentage change in the quality of service and β is a coefficient.

This approach seems to follow the standard price cap regulation model (see for instance, Armstrong, Cowan, & Vickers, 1994; Weyman-Jones, 2003). A financial plan agreed with the regulator forecasts, for an initial period of generally 5 years, operating costs (OPEX), investments (CAPEX), depreciation (D) based on the recognized RAB (regulatory asset base),[7] the planned increase of productivity (ΔX) and the level of profits to be recognized to the company, based on the WACC. The initial tariff T_0 is then calculated to assure that, also taking into consideration forecast tariff adjustments, the present value of forecast revenues be equal to the present value of costs and target profits:

$$\mathrm{T_0 = VA(OPEX + D + WACC \times RAB)/VA[(1 + \Delta P - \Delta X) \times Q]} \tag{3}$$

The core of price cap regulation is to allow the operator to keep profits resulting from having achieved productivity gains greater than forecast. However, at the end of the regulatory period there should be a 'claw back' of extra profits: in the following period all forecasts should be made anew, and the new tariff should be set so as to reduce forecast profitability to the level deemed appropriate by the regulator.

Actual regulation of highways tariffs in Italy is far from this model, in spite of the reference to 'price cap' (see Ragazzi, 2004).

There is one major aspect in the interpretation of the X parameter. This is set by ANAS with regard to a number of considerations, in addition to the expected increase of productivity: depreciation of planned investments, forecast traffic increase, compensation for past differences between planned and actual inflation, profitability to be recognized to the operator. Bundling together such different aspects reduces transparency and leaves a wide discretionality to ANAS in negotiating tariff adjustments with each concessionaire. This became evident when Autostrade's tariffs had to be renegotiated for the second 5-year period, at the end of 2002. An expert advisory committee of the Minister of the Economy estimated that tariffs should be increased by substantially less than had been agreed by ANAS: the difference, at the end of the period, was about 20%.[8]

The reasons for such wide differences were several, but they all boiled down to the very 'philosophy' of the price cap, and in particular to whether the 'claw back' of extra profits should be or not an essential feature of our regulatory system. Autostrade's case is relevant not only because it is the dominant operator but also because rulings applied to Autostrade are then reflected in the regulation of the other 23 concessionaires.

There is no doubt that Autostrade realized large extra profits compared to the original financial plan for the period 1998–2002. Revenues in the last year were 25% higher than forecast, ROI increased from 6.8% in 1997 to some 16%, net profits more than doubled. The reasons for this were essentially two: the increase of traffic, which was 22% compared to the 11% forecasted, and the volume of investments which barely reached 40% of what it had been originally envisaged, in part due also to administrative delays in approval.

It does not seem appropriate to pass on the 'traffic risk' to a licensee company: traffic growth depends from the expansion of the economy, the price of gasoline and other factors entirely outside the control of the operator. In an optimal risk allocation, the operator should assume only the (small) portion of traffic risk that depends on his or her leeway for stimulating additional traffic. Productivity measured as traffic volume per kilometre cannot obviously be taken as a measure of greater efficiency of the operator, in a price cap system.[9]

A different approach might be justified in the case of a new link to be built in project financing, where the 'traffic risk' is one of the variables that the promoter/operator has to consider in determining the profitability of the project. Instead, for long established networks, the 'traffic risk' depends essentially from the forecast assumed by the regulator in the financial plan on the basis of which tariffs are set. If the regulator makes a 'pessimistic' forecast the operator stands to obtain large extra profits, and vice versa. This is not a wise policy: it increases pressures to 'capture' the regulator, while the operator may risk to go bankrupt if the regulator is too strict or simply too optimistic in projecting traffic growth.

As we mentioned, up to the mid-1990ies government rules in Italy foresaw that traffic increases, net of the relative additional costs, should be reflected in a correspondent reduction of the tariff. The different approach followed by ANAS in adjusting Autostrade's tariffs for its second regulatory period was unsuccessfully contested by NARS; one can wonder what ANAS would have ruled, had traffic increased by less than forecast!

Concerning new investments, if their volume is less than forecasted in the original financial plan, while the tariff was set to include their amortization costs, the operator obviously benefits from extra profits. This was deemed acceptable by ANAS, due to the fact that planned investments were still to be realized, even if with long delays. For the second regulatory period (2003–2012) it was, however, agreed that tariff increases for amortization of new investments (spread over a 10-year period) would be recognized only when investments were actually underway.

Investment costs reflected in the tariff are not the actual costs born by the operator, but rather the amount agreed and 'negotiated' with ANAS. The operator stands to gain (or lose) for any difference between negotiated and actual investment costs. The operator might however refuse to carry on investments if the negotiated value is not acceptable to him: the cost reflected in the tariff is therefore likely to be higher than would have been if determined through a public tender for works.

The last term of the price cap formula, $\beta\Delta Q$, is supposed to measure improvements in the quality of service. Tariff increase due to this parameter is limited to a maximum of 0.75% per annum. As it is applied in Italy, Q is measured as the weighted average of two parameters: the quality of road pavement (60%) and the amount of accidents (40%).[10]

Regarding road pavement, it is not easy to distinguish between normal maintenance and new investments, since regulation is based on operators' balance sheets, which are not always clear on this point. There appears to be the risk that operators be remunerated twice for road pavement expenditures: for the resulting quality improvement and for amortization of new investments.

Another critical point is that the rate of accidents is substantially outside the control of the operator. Accidents are essentially a function of average speed.[11] In turn, average speed may be effectively limited only by police regulations. Stricter police enforcement of speed limits thus translates in a tariff increase, which seems unjustified. There is little that operators can do on their own to reduce accidents. In recent years, due to the introduction of higher penalties for exceeding speed limits, accidents declined, but more so on ordinary roads than on highways.

Operators also have a dubious interest in reducing average speed, since users may opt for ordinary free roads if they cannot achieve high speed on highways.[12]

The system applied in Italy is only nominally a price cap: there is no 'claw back' of profits and profitability is not limited to a target rate of return. Instead, the price cap is intended as a mechanism to determine tariff increases on the basis of the various parameters indicated above, with little or no consideration for the concessionaire's level of (extra) profits. This, and ANAS' generous application of the price cap formula, explains the explosion of Autostrade's and other concessionaires' profits and market values, over the last 5 years.

TARIFF POLICY AND EUROPEAN DIRECTIVES

The European Commission presented on 29 August 2003 a proposal for amending directive 1999/62/CE. The most innovative aspect is the goal of reducing external costs by differentiating tolls in order to reduce congestion and to charge costs according to vehicle characteristics. Appropriate charges could encourage cleaner engines and reduce the total number of lorries by increasing their size.

In Italy different fares are currently applied to different highways, but these differences reflect only historical and prospective costs. No attempt is made to relate the level of tolls to congestion, and there is no difference in charges according to the hour of travel. Vehicles pay according to the number of axles and not according to the weight or the pollution they cause.

Concessionaires are opposing any change away from flat tolling based on costs and conventions' financial plans. AISCAT (the Italian association of toll motorways companies) and ASECAP (the European association) justified the opposition to differentiated tolling by alleged technical difficulties (and costs); but one may think that a more fundamental reason for their opposition is the realization that differentiating tolls according to congestion would require a strong public agency to control the system, would much reduce their power to negotiate tariff increases and would expose to the public the perception that tolls are really a form of taxation and not the price for a service. Their prestige as entrepreneurs would be much reduced, and separating toll revenues from costs could open the door to 'unbundling', i.e. assigning through tenders the various services now all in the hands of the same concessionaire, which would be the death sentence for the nature of their business.

Be as it may, both organizations opposed the creation of independent authorities to oversee concessions, and argued that the European directive should not include highways under concession. In April 2004, the European Parliament approved an article which excluded highways under concession from the amended directive. The vote obtained a large majority, mainly thank to the Italian and French delegates. The European Council has then recognized the impossibility to find an agreement between the member states on a common policy for motorways under concession.

When companies' profits depend entirely on regulation, as in this sector, they wish to have a free hand in lobbying with politicians, and politicians wish to keep the regulating power in their hands.

NOTES

1. Autostrade was regarded as an efficient company, and achieving greater efficiency did not appear a relevant objective for privatization. It is not clear if efficiency has since increased. In the 3 years after privatization, the number of employees decreased by 10%, due mainly to the diffusion of electronic collecting systems, but this process had already started years before, with a reduction of 8% from 1995 to 1999.

2. ANAS also maintains some 20,000 km of state roads; provincial roads are about 145,000 km.

3. Austostrada del Brennero has interests also in the rail sector. The company, which is highly profitable, has been authorized to create a tax free fund to finance the new Brenner rail tunnel. This fund amounted to 232 million euros at the end of 2003, to be invested in treasury bonds. A subsidiary offers also international rail transport for freight (STR Brennero Trasporto Rotaia spa).

4. Selection of projects largely reflects political and local pressures, rather than traffic flows. Highways congestion is actually concentrated in northern Italy, around the most important cities.

5. For instance, Autobrennero claimed almost 550 million euros of credits towards ANAS alleging that its tolls had not been increased from 1991 to 1999. ANAS

proposed to offset this debt against an extension of the concession, from 2005 to 2014. The European Commission objected to such renewal without a tender, but ANAS justified it as necessary to recover the cost of past investments. If a different company wins the auction for renewal in 2014 it should pay back the investments made by Autobrennero between 2005 and 2014.

6. Planned inflation is set every year in the government's Economic and Financial Planning Documents (DPEF). However, following a controversial interpretation of Autostrade's concession contract, ANAS agreed to allow differences between planned and actual inflation to be recovered in tariffs, after the concessions for the first 5 years.

7. RAB is calculated annually as follows: $RAB_1 = RAB_0 + CAPEX_1 \ D_1$.

8. The Minister of Transport backed the ANAS, and Autostrade got the tariff adjustment as proposed by ANAS. To avoid interministerial conflicts, the tariff adjustment was approved by law, and the same law (47/2004) even decided that the X parameter should be revised after 10 instead of 5 years.

9. If productivity is defined as operating costs per standard traffic unit, productivity of Italian operators seems not to have increased appreciably over the last decade.

10. The roughness index is measured by cars equipped with special instruments directly by the licensees and then the results of the tests are then examined by ANAS. Accidents are registered by the police and compared with the volume of traffic.

11. Accidents are related mainly to two factors: speed and traffic. According to a study by the National Committee for Research (CNR), the number of deaths is proportional to the fourth power of the average speed of the traffic flow. The number of seriously injured persons increases with the third power of speed, the total number of persons injured increases with the square of the speed while the number of accidents increases by 2% for an increase in speed of 1 km/h.

12. An interesting example is the company Milano Mare–Milano Tangenziali, which in 1999 introduced very low speed limits in the urban tracts of the highway: 50 km/h for lorries and 90 km/h for cars. While accidents declined, congestion increased substantially. Under strong pressures from the transport industry, only 20 days later the speed limit for lorries was upgraded to 70 km/h and enforced only during daytime.

REFERENCES

Armstrong, M., Cowan, S., & Vickers, J. (1994). *Regulatory Reform: Economic Analysis and British Experience*. Cambridge, MA: The MIT Press.

Baldassarri, M., Macchiati, A., & Piacentino, D. (1997). *The Privatization of Public Utilities: The Case of Italy*. London: MacMillan.

Boitani, A., & Petretto, A. (1999). Privatizzazione e autorità di regolazione dei servizi di pubblica utilità: un'analisi economica. *Politica Economica*, *3*, 271–308.

Iozzi, A. (2002). La riforma della regolamentazione nel settore autostradale. *Economia Pubblica*, *4*, 71–93.

Ragazzi, G. (2004). Politiche per la regolazione del settore autostradale e il finanziamento delle infrastrutture. *Economia Pubblica*, *4*, 5–34.

Weyman-Jones, T. (2003). Regulating prices and profits. In: D. Parker & D. Saal (Eds), *International handbook on privatization*. Cheltenham, UK: Edward Elgar.

THE CONCESSION THROUGH A BID – THE NEW BRESCIA–MILAN HIGHWAY: A CASE STUDY

Fabio Torta

INTRODUCTION

The approximately 80-km, three-lane, tolled A4 motorway link connecting Milan with Brescia and Bergamo is subject to systematic congestion as is the whole road network. The corridor is one of the most industrialised and densely populated regions in the north of Italy.

The chambers of commerce of these three counties took the lead in sponsoring investment to increase capacity in the road corridor. Later, after the tailoring of quite complex technical and functional investment schemes, Brebemi SpA (a Ltd company whose shareholders include provinces and main motorways concessionaires) was established in 1999 as the promoter of a build, operate and transfer scheme in accordance to the new legislation for public procurement and project financing in Italy.

The feasibility of the project was submitted to four steps of assessment:

1. *Pre-feasibility study* (1997). Cost–benefit analysis of four investment alternatives: a minimum alternative for on-site widening of A4 (fourth lane); two options involving both substantial amount of tunnelling works along the existing infrastructure; a new motorway link connecting Brescia and Milan with a different layout than A4.

Procurement and Financing of Motorways in Europe
Research in Transportation Economics, Volume 15, 135–143

ISSN: 0739-8859/doi:10.1016/S0739-8859(05)15011-2

2. *Optimisation* (1998). Project variants identified in the pre-feasibility study (both involving the construction of a new motorway link) were further assessed to compare costs and benefits of alternative options concerning both routing and construction standards.
3. *Project approval.* In 2001, Brebemi SpA submitted for approval to ANAS (national company deputy of the Ministry of Public Works with reference to motorway concessions) a preliminary project scheme, with the new Brescia–Milano motorway link integrated with the southeast section of the road system bypassing the metropolitan area of Milan.
4. *Final project.* Further to the approval of the project-financing scheme as eligible for public funding, on January 2002 the Ministry launched an international tender to award the construction and the operation of the new motorway link.

In April 2003, Brebemi SpA was identified as provisional concessionaire and the definitive concession was granted in 2004. In March 2005, EIA Commission of Environmental Ministry approved the project. The definite design will be carried out before the end of 2005.

THE CONCESSION

The procedure followed the existing legislation at the start of the bid, the Merloni ter (109/94 and subsequent modifications). The Merloni law has completely redrawn the public works concession procedure and has defined the project financing rules, previously absent in the Italian legislation.

First, the modification of its juridical nature is confirmed according to EU regulations: the concession becomes a contract, not an administrative measure. Second, it states that the objective of the contract is constituted by:

- financing of works;
- definite and executive design;
- works execution;
- infrastructure management, which is the right to organize and sell the public service to people paying a toll.

During bid development the law was changed (the new version is defined as Merloni quater), with some modifications about the concession obligations and the same bid procedures, but the previous version of the law was applied to this procedure as it was already started.

The new law is facing some problems with regard to competition because it provides pre-emption for the promoter; consequently, the risk could be a reduced participation of several competitors. The public financing (the new version admits a share higher than 50%) and the concession period (now even longer than 30 years) should represent minor problems, unless past negative experience of the highway national concessionaires are considered (when the concession was still an administrative act): public contribution and concession period in many cases have largely exceeded these limits.

BID PROCEDURE (ACCORDING TO THE CONCEDENT INTERPRETATION – AUCTION VERSUS NEGOTIATION)

In the first phase, without promoter pre-emption, the law provides a bid to select two competitors vs. the promoter. This bid is organized according to the most economically advantageous proposal criteria (evaluation through a mix of technical and economic parameters), having the promoter's design as the bid base.

Quantitative and Qualitative Parameters Adopted for the Evaluation

Parameters and relative weights were pre-defined by ANAS to determine the most economically advantageous proposal. In this specific case, the parameters and the weights were as follows:

- operation procedure 20 points;
- technical–esthetical evaluation 12 points;
- toll level for users 18 points;
- concession period 16 points;
- construction period 15 points;
- economic return (operative costs/charge revenues average ratio, with reference to concession period) 10 points;
- sub-concession transfers 9 points.

Among these parameters there is no public subsidy reduction, because in the promoter PF scheme public contribution was not requested. On the contrary, the economic return parameter partially reproduces the toll-level parameter, because the operative costs are fixed for the part related to ordinary maintenance (minimum established by ANAS).

In this case, there were only two competitors and therefore the bid defined only the improvement of the competitors' proposal with reference to the promoter's design with respect to the indicated parameters. After the selection phase of the competitors, the law requires the negotiation between the promoter and the two competitors. This did not occur because ANAS, the grantor, organized an auction with three subsequent reserved raises and with respect to the following parameters (each indicated with its relative weight):

- toll level 18 points;
- concession period 16 points;
- construction period 15 points;
- economic return 10 points;
- sub-concession transfers 9 points.

The selected parameters were the same as those used in the first phase (selection of the competitors' proposals), and with the same relative weights with the exception of the qualitative parameters (technical and esthetical evaluation and operation procedure). The selection of only quantitative parameters strengthens the auction concept in comparison to the negotiation concept.

Certainly, this type of bid includes qualitative and trust aspects as well. Consequently, the public decision-maker has to be strong and have competence and skill to manage technically and economically the discussions with the promoter and the competitors. Some discretion in the evaluation and selection process of the winner is indeed inevitable. It is probable that the ability to negotiate could develop through experience and time. This bid to assign a concession of a completely new highway infrastructure was the first and unique in Italy; therefore, the auction choice was probably unavoidable, also considering the guarantee in terms of transparency.

On the other hand, the negotiation could have spared a few doubtful results from a technical standpoint. The negotiation could have also forced the grantor and the competitors to concentrate on more interesting and qualifying parameters, perhaps covering also the designing aspects.

In particular, the negotiation could have avoided the competitors' three raisings on some parameters:

- construction period, reached to a fantastic time of 31 months (not impossible, but objectively questionable);

- economic return that, selecting the participants with adequate guarantees, dimensions, technical capabilities, etc., is not simply reducible, unless running the risk of financial feasibility, having simultaneously to guarantee good managing levels;
- the transfer to the grantor (ANAS) of a sub-concession revenues quote, a parameter that shows some conflict of interests of the decision-maker.

BID RESULTS: CONSEQUENCES ON THE CLIENTS AND THE COMMUNITY, ON THE GRANTOR, AND ON THE CONCESSIONAIRE

Tables 1 and 2 show the bid consequences:

Clients / Community

- Lower tariffs. With reference to the promoter's proposal the tariff level has been reduced by 16%.
- Better performances, but with some problems, particularly on the structures whereas quality becomes aesthetics (generally the beauty costs) and on some environmental interventions: the so-called design quality, a rather qualitative parameter, is objectively fragile in evaluation procedures that include more robust quantitative parameters.

Table 1. First-Phase Bid – Competitors' Design Selection.

Elements of Proposal Evaluation	Basis Values of Tender (Promoter Design)	Values Resulting from First phase of Tender (Best Competitor's Design)
Tariff level (Euro/km)		
Light vehicles	0.06998	0.0657812
Heavy vehicles	0.12443	0.1169642
Concession period (years)	30	25.5
Construction period (months)	48	42
Return[a]	20.70%	15.96%
Sub-concession rights to the grantor	2%	10%

[a]Operative costs/charge revenues average ratio, with reference to concession period.

Table 2. Second-Phase Bid – Concessionaire Selection.

Elements of Proposal Evaluation	Basis Values of Tender (First Phase)	Values Resulting from Second Phase Tender (Negotiation) – Last Auction Raising					
		BREBEMI		ASTALDI		GEFIP	
		Proposal	Points	Proposal	Points	Proposal	Points
Tariff level (Euro/km)							
Light vehicles	0.0657812	0.05865	18.00	0.0615338	0.04	0.06074	4.98
Heavy vehicles	0.1169642	0.10428		0.1094091		0.108	
Concession period (years)	25.5	19.5	16.00	21.96	0.03	19.67	14.93
Construction period (months)	42	31	15.00	37.97	0.07	34	8.57
Return[a]	15.96%	15.18%	10.00	15.55%	0.98	15.54%	1.22
Sub-concession rights to the grantor	10%	40%	9.00	28%	0.00	31%	2.25

[a]Operative costs/charge revenues average ratio, with reference to concession period.

Grantor

- The concession period is shorter (from 30 to 19.5 years).
- The revenues derived by sub-concession rights increase.

Concessionaire

- The management of the procedures (after the bid) such as approval, local agreement or the means to acquire, for example, land, is feasible in a weak condition: the timing function (31 months) includes these procedures and cannot be changed because of bid and financial constraints.
- There is the obliged but positive need to increase the efficiency of the building and operative procedures.
- The bid, as it seems from the results, should have already optimized the productivity. The applied convention with the grantor includes a price-cap mechanism. The formula applied to the other concessionaires includes the productivity (equal to 0 for the first 5 years) while the variables related to the quality include the characteristics of the road surface and the number of accidents (both should improve). The characteristics of this price-cap underestimate that this highway is completely new (to integrally build, without links with an existent network, a real PF, separated from other factors) with a recent high standard design. The present price-cap awards subjects who have a low service level (e.g. the history of the restructuring of the Milan–Turin highway or the Rome–Teramo highway, to be re-assigned after the loss of the previous concession, which was won with a very low tariff, although aware of the low quality at that time and the forecast of rapid increase toll charges) and not subjects who start with high levels, or who win a bid building inclusive. In the case of a new highway, with design qualitative standards, at least initially included in the bid procedures, it is not clear whether the mechanism of a tariff incentive helps to reach a higher standard: it seems enough to fix maintenance technical parameters and to control adequately the activities that are carried out.

With reference to the weight scheme of the previous parameters, other consequences are appraisable. In particular, we can estimate, with some simplification (on the basis of the preliminary estimated traffic revenues), the order of magnitude of the concessionaire cost to obtain a single point:

- tariff level — 10–15 million Euros;
- concession period (year) — 30–40 million Euros;
- construction period (month) — 2–3 million Euros.

We can observe that the construction period parameter is probably overestimated when compared with others, particularly considering the real benefit for the community: for example, the delivery of the infrastructure 2 months in advance will represent 2–3% of the total time to complete the investment from the bid to the start of operation.

Specific Problems

Project Modifications after the Bid

The assignment on the basis of the preliminary project must take into account all the approval procedures and the modifications in the environment sector (EIA) or related to the assent (territorial requests).

It is important to point out that the law admits the possibility to revise the concession, also through the prorogation of the concession period. All the variations to the assumptions and the basis conditions, derived from the grantor administration decisions, from new rules or laws that modify the pricing mechanism or from management procedures, if affecting the economic–financial equilibrium, require the revision of concession conditions.

The hypothesis of concession revisions is admitted also in case of modifications in favour of the concessionaire; this aims at avoiding extra profit situations.

Apart from the formal aspect it would be interesting to monitor the differences of the project, of the amount, changes of tariffs or of concession period between the values defined through the bid procedures and those effectively realized.

Apparently, the approval procedures during 2004 and 2005, the new law on structure dimensions and local requests are causing considerable modifications in the design resulting in a significant increase of investment costs in the Brescia–Milano highway project.

Grantor and Planning Institution Role

The national choices about concessions that have an impact on restricted areas lead to difficulties in dialog at the local level (region, provinces and municipalities). The management of the preliminary phases is complicated: the promoter does not play a definite role until the formalisation is completed; the dialog between the grantor institution (ANAS) and the principal planning institution (Region) does not develop along strategic planning aspects, but is possible only on a single and definite project.

Local Planning and Pricing

Local ordinances and directives, after the bid procedures, can invalidate some results obtained through the bid: changes in route layout, environmental modifications and different structures often tend to worsen the economic–financial framework. The single PF scheme, on the contrary, could not optimize a network scenario, whose management should be opportunely unitary.

The lack of public resources, when these types of interventions impact on the economic balance of the concessionaire, prevents substantially a widening of this aspect. Even local requests for a free or discounted use of the infrastructure granted to the residents in the municipalities crossed by the highway, for new accesses to the highway, etc., are obtained with difficulty in a PF scheme, if it is not part of complete planning. In such cases the public institution has to consider hypotheses of direct contribution, in case it wants to maintain the governance on the traffic and territory.

FINANCING MOTORWAYS IN POLAND

Monika Bak and Jan Burnewicz

INTRODUCTION

The distinction between Poland and other European countries in the case of motorway network density has grown during the past decades. The underdevelopment of motorways and expressways causes an isolation of Poland from the rest of Europe. In spite of the acceleration of investment efforts in the last years, the difference between Poland and different European countries has still remained huge. Taking into consideration the length of motorways per 1 million citizens in 2000, Poland takes the 27th place in the list of 31 European countries. The indicator for Poland amounts to 10.3 km per 1 million citizens while the average number for 31 countries is 106.3 km per 1 million citizens. In order to achieve this European average, the length of motorways in Poland should increase from 523 to above 4,000 km, which would require an investment of more than 70 billion PLN (ca. €15–16 billion). The comparison of the development of the Polish motorway network with the conditions in other new EU member states is very unfavourable for Poland. In spite of the success in the process of transformation as well as positive legal and organisational changes influencing the implementation of investment plans, the pace of motorway construction from the 1990s onwards has still not been as dynamic as could be expected.

Procurement and Financing of Motorways in Europe
Research in Transportation Economics, Volume 15, 145–155

ISSN: 0739-8859/doi:10.1016/S0739-8859(05)15012-4

BRIEF HISTORY OF MOTORWAY DEVELOPMENT IN POLAND

After the Second World War until the beginning of the transformation period, only ca. 100 km of motorways were built in Poland. But the conceptions and programmes have been very ambitious. In post-war history, the first plan of motorway construction was dated 1946. In this project, a special emphasis was put on North–South connections. Projections were impressive, but they were not carried through. The next programme 'Model of Road Network in Poland' was developed in the 1960s under the auspices of the UNDP. Then, in the 1970s, when some spectacular infrastructure investments were realised (especially in building and heavy industries), several sections of motorways were also built (Silesia and Warsaw as preferred traffic nodes). In 1991, the first programme after the initiation of transformation was published. It assumed the realisation of 2,600 km of motorways, including the A-1 from Gdansk to the Czech border, A-2 from the German to the Belarussian border, A-3 from Szczecin to the Czech border, A-8 from Wroclaw to Lodz and A-12 from Olszyna to Krzyzowa (as a supplement to A-4). Further programmes, including the programme of motorway construction in Poland of 1993 and the subsequent one (Programme, 2003) were more realistic and assumed to build only three new motorways with a total length of 1,994 km, i.e. A-1 (564 km), A-2 (651 km) and A-4/A-18 (779 km).

PRESENT SITUATION OF MOTORWAY DEVELOPMENT

At the turn of the century, some positive results in motorway expansion in Poland appeared. Between 1946 and 1979, only 109 km of new motorways were built; 80 km were added in the 1980s and another 156 km in the 1990s. From 2000 to 2004, an additional 119 km were completed. Nevertheless, the government has still not managed to realise its plans, i.e. the building of 550 km of motorways in the years 2002–2005 (including 150 km of totally modernised sections of existing expressways) and the start of construction of further 500 km of new motorways (Infrastructure, 2002). The present condition of the motorway network in Poland is presented in Table 1.

Table 1. Realisation of Motorways Construction in Poland.

No.	Planned Route	Total Planned Length (km)	Operated Sections (km)	Charges	Main Difficulties
A-1	Gdansk–Torun–Lodz–Czestochowa–Gliwice–Gorzyczki (on the Czech border)	564	17.5	Toll-free	Delays of starting construction within concession systems for section Gdansk–Torun, new investments predicted start in 2005
A-2	Swiecko (border with Germany)–Poznan–Warsaw–Siedlce–Terespol (on the Belarussian border)	651	146	13 km – toll-free; 135 km – toll	Western part under construction within concession system; delays in realisation of the eastern part (from Warsaw to the Belarussian border)
A4	Jedrzychowice (on the German border)–Wroclaw–Katowice–Cracow–Korczowa (on the Ukrainian border)	779	350	61 km – paid/toll; 289 km – toll-free	Under construction, mainly public funds used, paid section operated by concessionaire Stalexport
A-6	Kolbaskowo (on the German border)–Szczecin	21	14	Toll-free	Modernisation of remaining section planned
A1-8	Olszyna (on the German border)–Krzyzowa (to A-4)	70	0	Toll-free	Modernisation from 2005 onwards

EXPERIENCES – DIFFERENT SCHEMES OF FINANCING

Until 1993, only traditional solutions based on budget funds were used in motorway financing. While approving the programme of motorway construction in 1993, the government assumed that the state budget would be able to cover only 10–15% of the total costs of investments and that the development of a concession system of financing would expand (Programme, 1993). In reality, it turned out that from 1994 to 2001 the share of private concessionaires was minimal.

The trial of the diversification of financial sources was the formation of the National Motorways Fund, which was allowed to apply for financing from the following sources: loans from international financial institutions, revenues from the sale of vignettes, emission of bonds secured by replacing future ownership rights or state budget shares, emission of bonds guaranteed by the state, secure instruments concerning future revenues (e.g. from vignette sale) and the sale of property assigned by the state treasury. Afterwards, the plan to replace toll motorways by the vignette system arose (vignettes were introduced for freight road transport in January 2002). But the introduction of common use vignettes, also for passenger cars, was cancelled in spring 2003 due to political and social resistance.

As mentioned above, in the programme of 1993, the government also assumed that motorway construction in Poland could be realised with the help of the concession system. Due to poor results in the realisation of the project, the pure BOT system financing was later transformed into public and private partnerships setting up an increase of state funds involvement. A summary of the construction and operation of motorways in the concession system is presented in Fig. 1.

As shown in Fig. 1, only one concessionaire in Poland has true practice in the range of motorway construction, i.e. Autostrada Wielkopolska S.A. The second one has not yet started construction activities since a contract between GTC and the government was only signed in 2004, though the concession had been granted in 1997 (delays resulted from the lack of agreement concerning the price of the motorway as well as from financial difficulties of the concessionaire). The third concession, Stalexport Autostrada Małopolska S.A., was only granted the right to adjust the charging system and to operate a 61-km long section of the A-4 motorway from Katowice to Cracow.

Autostrada Wielkopolska S.A. has been developing the biggest project in Poland with regard to the concession system and structure, according to the PPP programme. The company was founded in January 1993. In order to meet its obligations under the Concession Agreement, the Development Company was established — A2 Bau Development GmbH (founded by the shareholders of AWSA: Strabag AG and NCC International AB) which is responsible for the construction. In addition, an operating company was founded – Autostrada Eksploatacja S.A. (founded by the shareholders of AWSA: Transroute International S.A., Kulczyk Holding S.A. and Strabag AG). In order to ensure proper performance under the contracts, in strict

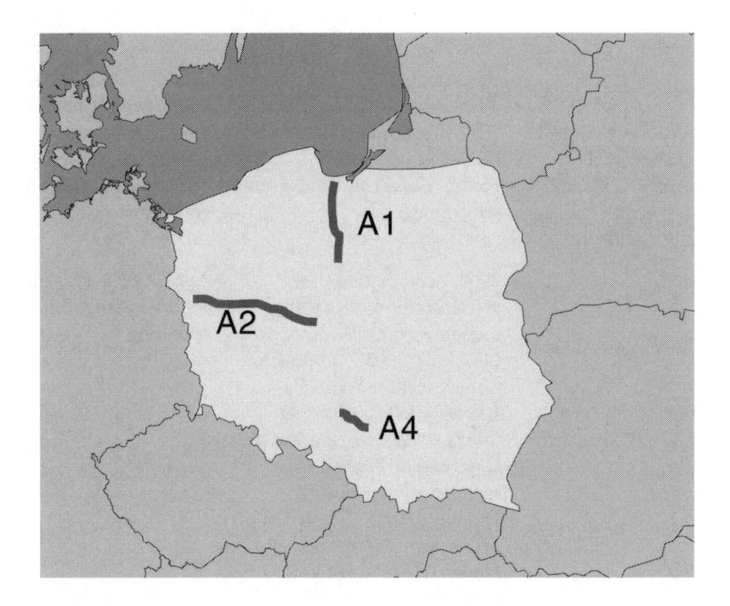

Fig. 1. Summary of the Construction and Operation of Motorways in the Concession System

A-1	A-2	A-4
Concession subject		
Construction and operation of motorway Gdansk–Torun, 152 km	Construction and operation of two sections of motorways: Konin–Nowy Tomyśl (including modernisation of existing section Konin–Wrzesnia), without ring-road of Poznan built by public sources; and Nowy Tomysl–Swiecko	Adjusting to charging system and operation of section of A-4 motorway from Katowice to Cracow (61 km)
Concessionaire		
Gdansk Transport Company	AWSA (Autostrada Wielkopolska S.A.)	Stalexport (from 2004 Stalexport Autostrada Małopolska S.A.)
History of granting and concession and conclusion of a contract		
December 1995 – Agency of Motorways Construction and Operation (ABiEA)	September 1995 – Agency of Motorways Construction and Operation (ABiEA)	June 1995 – Agency of Motorways Construction and Operation (ABiEA)

A-1	A-2	A-4
advertised for tenders; 1997 – granting concession; August 2004 – conclusion of a contract between GTC and government	advertised for tenders; March 1997 – granting concession (for 40 years), conclusion of a contract between AWSA and government	advertised for tenders; March 1997 – granting concession conclusion of a contract between Stalexport and government
Realisation of a contract		
No	2002 – completion of modernisation of existing section Konin–Wrzesnia (47.7 km); 2003 – completion of new section Wrzesnia–Krzesiny (37.5 km); 2003 – taking over for operation a ring-road of Poznan (free segment of 13.3 km); 27 October 2004 – completion of new section Komorniki–Nowy Tomysl (50.5 km); remaining section – Nowy Tomysl–Swiecko (104.5 km) – planned completion in 2007	April 2000 – charging of fee, concessionaire responsible for operation
Charges		
No	Passenger cars and motorcycles – ca. €0.02 per 1 km; trucks, two axles – ca. €0.032 per 1 km; trucks, three axles – €0.046 per 1 km; trucks >3 axles – €0.076 per 1 km; non-standard vehicles – €0.196 per 1 km	Passenger cars and motorcycles – ca. €0.02 per 1 km; trucks – ca. €0.043 per 1 km

compliance with Polish law and the provisions of the Project Agreements, the Parties to the concession, the Minister and the Concessionaire, appointed the independent engineer – WS Atkins of the UK – whose duty has been to supervise the design process, the construction and operation of the motorway as well as to care for the proper execution and adequate quality of the work.

The shareholders of Autostrada Wielkopolska S.A. are: Polskie Sieci Elektroenergetyczne (19.7%), Kulczyk Holding (14.9%), Bank Zachodni WBK (5.4%), PKN Orlen (9.2%), TUiR Warta (4.8%), Egis Projects (3%),

Strabag AG (10%), NCC International AG (10%), Kulczyk Investment GmbH (19.1%) and others (3.9%). Polish investors amount to 57.9% and foreign investors to 42.1%. The financing of the motorway was based on the Sponsors' Equity (more than 27%), long maturity loans from the European Investment Bank (33%) and long maturity commercial bank loans (26%). The remaining 15% comes from toll revenues and the interest on balances during the construction period (for more information, see, e.g. the official website of the company).

In 2002, the government made another attempt to stimulate the investment in motorways. It planned to intensify public activities by implementing a programme and to take over the initiative in the diversification of sources of financing as well as in the efficient start up of new projects. In order to achieve that purpose, the organisational structure responsible for motorway construction was strengthened. Therefore, the Agency of Motorways Construction and Operation was joined with the General Directorate of Public Roads to form the General Directorate of National Roads and Motorways. This new institution is obliged to work efficiently on the development of motorways as well as expressways and remaining public roads. Simultaneously, however, this did not mean that the extension of state responsibilities led to an abandonment of private partnerships.

By 2003, still no radical progress had been made. Again, the crux of the matter consisted of producing additional sources of financing. In order to break an impasse, the government suggested several initiatives in the legal field:

1. Act of specific rules of preparation and realisation of investments in the scope of national roads (approved by the Parliament/Seym, 10 April 2003), which considerably simplified the procedure of locating and obtaining real estates in the investment process.
2. Act amending an act of paid motorways of 1994, with a major provision to form additional financing instruments in order to support road network managed by the General Directorate of National Roads and Motorways. On the basis of the Act, the previous National Motorways Fund (in practice not implemented) was replaced by the National Road Fund, supplied by a new fuel charge. The special fuel charge is the latest proposal on financing Polish motorways (introduced in 2004, on the level of 105 PLN (€23.3) per 1 ton of fuel, valorised annually according to the rate of inflation). Total revenues in 2004 were ca. €240 million. That is ca. 6.5% of the total fuel taxes revenues, which amount to €3,685 million (Central Statistical Office, 2004).

PLANS FOR THE FUTURE – THE ROLE OF EU FUNDS

The Polish membership in the EU since May 2004 radically increased the amount of financial means available for realising the programme of motorway construction in Poland, though experts from the European Commissions sought to prioritise investments in the railway network. According to the settlement of 2004, the amount of EU assistance in the years 2004–2006 concerning the realisation of investments in three programmes (established especially for this purpose: (A) Sectoral Operational Programme– Transport, (B) Strategy of Use Cohesion Fund in Transport, and (C) Integrated Operational Programme of Regional Development will amount to €4236.1 million. Additionally, it is expected to devour €692.3 million of the state budget means, €504.3 million of self-governments, and €13.5 million of private investors' means. In total, it will amount to €5.4 billion.

The strategy of using structural funds and the Cohesion Fund in the transport sector should guarantee positive economic, social and ecological effects (if the financial means are used efficiently). Moreover, the strategy assumes to realise the concept which is coherent both internally and with the EU transport system. Then, within the SPO-T programme, initiatives such as the concentration of investment in transport corridors are treated preferentially.

For the sake of the weak competitiveness of Polish economy and the low level of GDP per capita in relation to the average indicator of the EU, in the years 2004–2006, all Polish voivodships have been classified to the area of Objective 1 of the structural policy of the EU. Financial means from ERDF in the transport sector will be distributed in Poland on the basis of two crucial documents, i.e. Sectoral Operational Programme as well as Transport and Integrated Operational Programme of Regional Development (Table 2).

The year 2004 happened to be a turning point in the intensity of financing in Polish transport infrastructure: the annual level of expenditure will rise from €1.5 billion in 2002 to up to €3.5 in 2006, which means the increase of transport infrastructure investments with relation to GDP will be up to 1.8%. For many years this level seemed to be unattainable. Now there exists a real chance to maximally profit from EU financial assistance, also by reason of preparing high-quality three structural programmes.

Thanks to EU co-financing, until the end of 2008, it will be possible to build over 250 km of motorways and 81 km of expressways. In spite of the extended investments, after 2008, the density of Polish road network will have only slightly improved in relation to the EU average. But the situation

Table 2. Co-Financing of Transport Projects in the Years 2004–2006 by Use of Structural Funds of the EU, in Million €, as of 2004.

Activity	EU Funds	State Budget	Self-Governments	Private Means	Total Public	Total Public and Private	Expenditure per 1 km in Million €	Output Indicator[a] (km)
I. Projects of sectoral operational programme – Transport – SPOT (ERDF)								
1.1.1. Railway line Warsaw–Lodz	231.5	77.1	0.0	0.0	308.6	308.6	3.0	102.1
1.1.2. Passenger railway fleet	51.1	17.1	0.0	0.0	68.2	68.2		30
1.2. Infrastructure access to seaports	119.8	23.7	16.2	0.0	159.7	159.7		NA
1.3.1. Logistic centres	13.0	4.3	0.0	6.5	17.3	23.8		1.0
1.3.2. Multimodal terminals	10.7	3.6	0.0	5.4	14.3	19.7		5.0
2.1.1. Motorways (including 2 Skierniewice–Warsaw)	299.8	100.3	0.0	0.0	400.1	400.1	5.5	72.6
2.1.2. Expressways (including S22 Elblag–Grzechotki)	56.8	18.9	0.0	0.0	75.7	75.7	1.5	51
2.1.3. Rebuilding of national roads and ring-roads	190.7	63.6	0.0	0.0	254.2	254.2	3.0	84.3
2.2. Modernisation of national roads in poviats	164.0	0.0	54.7	0.0	218.7	218.7	3.5	62.5
2.3. Road safety	17.0	5.7	0.0	0.0	22.7	22.7		NA
3. Technical assistance	9.0	3.0	0.0		12.0	12.0		NA
Total SPOT	1163.4	317.3	70.9	11.9	1551.5	1563.4		

Table 2. (*Continued*)

Activity	EU Funds	State Budget	Self-Governments	Private Means	Total Public	Total Public and Private	Expenditure per 1 km in Million €	Output Indicator[a] (km)
II. Transport projects financed from cohesion fund								
1. Modernisation of railway lines	895.4	158.0	0.0	0.0	1053.4	1053.4	2.4	434.5
2. Motorways	934.2	165.2	0.0	0.0	1099.4	1099.4	6.0	182
3. Expressways	179.7	30.9	0.0	0.0	210.6	210.6	7.0	30
4. Rebuilding of national roads	80.0	14.7	0.0	0.0	94.7	94.7	1.0	95
Total cohesion fund	2089.3	368.8	0.0	0.0	2458.1	2458.1		
III. Integrated Operational Programme of Regional Development (ERDF)								
1.1. Modernisation and extension of regional systems	768.6	0.0	256.2	1.0	1024.8	1025.8	5.2	199
1.6. Public transport in agglomerations above 500 thousand citizens	167.9	0.0	167.9	0.0	335.8	335.8	14.6	23
3.1. Rural areas – construction and modernisation of gmina and poviat roads of local importance	46.9	6.3	9.4	0.6	62.5	63.1	0.3	215
Total IOPRR	983.4	6.3	433.5	1.6	1423.1	1424.7		
Total	4236.1	692.3	504.3	13.5	5432.8	5446.3		

Source: Documents, 2004.

[a]The output indicator is one of the indicators used in Documents – operational programmes prepared in each New Member State in a process of applying to EU structural and cohesion funds. The indicator means product in km, e.g.number of km of new roads.

should then be significantly ameliorated in some transit connections and congested segments of roads in agglomerations and other cities.

CONCLUSIONS

The economic and social development of Poland in the last one or two decades should result in the modernisation of transport infrastructure, especially in some regions characterised by either a scarce infrastructure network or by a high demand for new investments. The need for new motorways has clearly been identified. The growth of road traffic, resulting from the increased mobility of citizens, a boom of private motorisation as well as the dynamic development of freight transport in turn generates an increase in the demand for high-quality road infrastructure. Only such types of infrastructure can ensure short travel time and a high safety level. Moreover, the transformation processes favour the development of new financing instruments of infrastructure investments, including the involvement of private capital, a concession system and public–private partnerships. In post-socialist conditions, these new forms of financing should stimulate investments in the domain of motorways. Polish practice indicated that the conditions favoured did not mean a simple shift towards a dynamic evolution of motorways investments. In Poland, some legal, social and financing barriers have blocked the realisation of motorway investments, especially in the North–South axis. Nevertheless, hopefully present solutions and new possibilities as well as EU support can stimulate the realisation of motorway construction. It should result in a considerable improvement of traffic conditions in road transport in Poland.

REFERENCES

Central Statistical Office. (2004). *Statistical Yearbook*. Warsaw: Central Statistical Office.

Programme. (1993). *Motorways construction in Poland.* Ministry of Transport and Maritime Economy. General Directorate of Public Roads: Warsaw.

Programme. (2003). *Motorways construction in Poland of 2003*. General Directorate of National Roads and Motorways: Warsaw.

PORTUGUESE EXPERIENCE IN MOTORWAY CONCESSIONS WITH REAL AND SHADOW TOLLS

Carlos Fernandes and José M. Viegas

INTRODUCTION

Since the early 1970, Portugal has had an interesting history of using motorway concessions as an important instrument for the expansion of its motorway network, with varied forms of tendering and contracting with private partners, as well as with different forms of engaging revenues from users in the recovery of the construction, maintenance and operation costs of that infrastructure.

This chapter presents the main facts and comments on the reasons for the adoption of each solution as well as comments on its consequences. Greater emphasis is placed on the shadow toll motorway programme given its relative novelty in the international scene and the controversy it has generated domestically.

EARLY STAGES: BRISA

The Portuguese experience of engaging private agents in the construction, operation and management of motorways started when the first concession was tendered and a contract signed with BRISA – Auto Estradas de Portugal, S.A. (hereafter BRISA) in November 1972. BRISA was initially incorporated as a shareholding society with wholly private equity, of which

Procurement and Financing of Motorways in Europe
Research in Transportation Economics, Volume 15, 157–174

ISSN: 0739-8859/doi:10.1016/S0739-8859(05)15013-6

about 60% was from Portuguese banks. Following the revolution in 1974 and the nationalisation of the banking sector in 1975, a new situation was created in which the state became the majority shareholder of the concessionaire. Through a series of bilateral deals and equity increases, the state increased its share of equity to 89.7%, while two positions of 5% were held by two public entities.

In the initial contract, with the exception of a toll revenue guarantee in the network under operation, the state did not provide any other financial support to the concessionaire. The guarantee consisted of a disposition allocation: in periods of traffic above forecasts, 90% of the 'excess revenue' to a special fund and 10% for acquisition of BRISA shares by the state; in periods of traffic below forecasts, a transfer from the fund to the concessionaire while the fund lasted, followed by zero interest loans by the state to BRISA, to be repaid as a first priority as soon as the company became profitable.

In the meantime, the entry of Portugal in to the European Community (in 1986) and the internal political evolution led to a progressive reduction of direct engagement of the state in the economic activity. Thus, at the end of the 1990s, the state alienated all its shares in four consecutive privatisation phases for a total revenue equivalent to €1.875 billion (Tribunal de Contas, 2003).

The initial concession contract of BRISA included the A1 (Lisbon–Porto), A2 (Lisbon–Setúbal), A3 (Porto–Braga) and A5 (Lisbon–Cascais) motorways for a total extension of 390 km.The current concession contract includes 11 motorways for a total extension of 1,100 km.

In 1985, the state, as majority shareholder, committed itself to contribute to the financial balance of the concession through direct participation in the investments (engineering projects, expropriations, construction, equipments and complementary works) and issuing guarantees for the operations financed by the company. Initially, this participation was 40% of each investment, which was then reduced to 35% and in 1997 to 20%. EU funds directly received by BRISA must be subtracted from the values thus computed. The sum of these participations until the end of 2001 was €800 million (at current prices of the various years). When the concession was allocated BRISA was exempt from all taxes and charges owed to the state and local authorities. Since 1997, BRISA is no longer exempt from corporate tax, but can still deduct from the collectible up to 50% of reversible physical investments made between 1995 and 2002, in the part that does not include participation by the state. This deduction can be made in the values of corporate tax relative to the years between 1997 and 2007. The accumulated value of these fiscal benefits between 1997 and 2001 was €671 million at current prices of the various years (Tribunal de Contas, 2003).

In order to foster competition in the motorway operation sector, BRISA was forbidden by the government to bid for motorway concessions in the period between 1996 and 1999, during which two concessions (west and north) were tendered for a total of 253 + 170 km. Since 1999, BRISA has been authorised to compete directly or through subsidiaries for road concessions within the object of its statute, i.e. construction, maintenance and operation of roads or service areas, as well as to promote the study and construction of social facilities. Currently, the main sources of BRISA's financing are: (i) equity; (ii) state participation and (iii) bank debt. According to estimates of the company, the shareholder IRR is 6.81% at constant prices.

On balance, we can see that although BRISA started as and later returned to the status of a private company, during its long period as a government-owned company it introduced various schemes that are certainly more protective than the current standards for private companies. Of course, the value of these protections has been considered by the private buyers of the company in the initial privatisation, as well as in the subsequent transactions of its shares in the Portuguese stock exchange, where it remains one of the most traded titles.

THE SECOND TAGUS BRIDGE (VASCO DA GAMA BRIDGE)

The first experience of public–private partnership[1] under the form of project finance has been the concession of the two Lisbon Tagus bridges. Following an international tendering process, the Portuguese State and Lusoponte-Concessionária para a Travessia do Tejo, S.A. (hereafter Lusoponte) signed a public works concession contract in March 1995 for construction, maintenance and operation of the second (Vasco da Gama) bridge, and for maintenance and operation of the first (25 April) bridge. Lusoponte had a wholly private international equity structure.

In the initial contract, the end of concession would occur not later than 33 years after the signing of the contract, when the following conditions would all have been met:

- Integral payment of the loans received under the financing contracts of the concessionaire.
- Total traffic volume in the two crossings in both directions, counted from 1 January 1996 (date of transfer of the operation of the first bridge to the concessionaire), reaching 2,250 million vehicles.

The concessionaire has the right and duty to collect tolls in the two bridges, fully assuming the traffic risk.[2] The concession was granted in a regime of exclusivity for all Tagus road crossings downstream of the Vila Franca de Xira Bridge (some 20 km north of Lisbon).

The concession contract provided for a significant increase of the toll on the 25 April Bridge immediately after the beginning of the concession, in order to reduce the volume of government funds brought into the project. However, in 1994 (while still in preparation for the signing of the contract), the official announcement of this increase evoked a very strong public outcry and disarray, which led to only minor adjustments of the toll. This decision led the state to assume as its own cost the toll revenues no longer collected by the concessionaire. The contract for financial rebalance of the concessionaire (Global Agreement for Financial Rebalance) was signed in July 2000, and included a scheme for additional payments by the state, twice per year over a period of 19 years, and the definition of the term of the concession at the end of 33 years (for further details, see Lemos et al., 2004). Financing the Vasco da Gama Bridge was as follows (for a total of approximately €900 million):

- EU Cohesion Fund: €320 million (35%).
- European Investment Bank loan: €300 million (33%).
- Tolls collected on 25 April Bridge: €50 million (6%).
- Shareholders: €66 million (7.3%).
- Others (basically government support): €164 million (18.7%).

Out of the total cost of the project, about €644 million was for construction and the remainder for maintenance of crossings, expropriation, relocation and environmental protection measures.

In July 2000, with the Vasco da Gama Bridge in full operation since March 1998, simultaneously with the Global Agreement for Financial Rebalance, the concessionaire renegotiated its financing structure, which now has an important component of commercial bank loans.

PUBLIC–PRIVATE PARTNERSHIPS FOR MOTORWAYS SINCE 1996

National Road Plans and their Execution as Framework for the Policy

The National Road Plan currently in effect (hereafter PRN 2000) was approved in 1998, with later changes approved in 2003, and follows those approved in 1985 and 1945. Two networks are defined in this new plan: the fundamental network and the complementary network.

The fundamental network has nine principal itineraries (IPs), connecting the main cities among themselves, as well as with ports, airports and land borders. The complementary network consists of complementary itineraries, connecting secondary cities among themselves and with the IPs, as well as by national roads (ENs) and regional roads (ERs). In 1996, the fundamental network was 50% complete, while the complementary network still had 65% of its extension to be built.

Thus, a total of 2,250 km of sections integrated in the fundamental and complementary networks still had to be built (50% of which with motorway profile),[3] which corresponded to a total investment of about €14 billion, at 2002 prices, equivalent to 11% of the Portuguese GDP of that year.

Considering that the yearly budget of the Public Road Agency (then called JAE – Junta Autónoma das Estradas; later called IEP – Instituto das Estradas de Portugal, and recently renamed once more as EP – Estradas de Portugal EPE) would stay at 0.65% of the national GDP and that, respectively, 25 and 27% of that budget would be channelled to build IPs and ICs; the period needed to fully build the planned network was 30 years.

In 1997, the Portuguese government opted for launching a vast programme of public works concessions. Initially, two concessions for motorways with real tolls were launched (west and north concessions), then seven concessions with shadow tolls (called SCUT – Sem Custo para o Utilizador, i.e. without costs for the user), and later two more concessions with real tolls. For another tract called 'Litoral Centro', with a total of 93 km, the concession contract was signed in September 2004. This was the first concession contract won by BRISA in its new competitive statute.

Motorways with Real Tolls

The model underlying these concessions stated that the private sector would be responsible for conception, construction, financing, maintenance and operation of the motorways for a period of 30 years. In these motorways the state would participate in the construction and operation costs only as far as the payments from the users would not be sufficient to meet the remuneration of private investment. Thus, the amount of public subsidy would basically depend on the construction cost per kilometre, on the foreseen traffic volumes and on the length of the concession period, as the toll levels were fixed by government at the outset.

The west concession was the first to be contracted but this option for toll collection was shortly after suspended by the parliament based on the following arguments: (i) part of these roads had been designed, financed and built earlier as free access motorways; (ii) they had been partly financed by EU funds; (iii) citizens of these regions did not have adequate road or railway alternatives for their displacements and (iv) traffic characteristics were dominantly local and inter-regional. Toll collection was suspended in the new sections, and only reintroduced after completion of the upgrade works on the alternative free access road (EN8), which was paid by the state.

This process strongly influenced the policy of the government at the time, as many of the motorways foreseen in the National Road Plan and not yet built did not have adequate alternatives, or were even expected to be built on the platform of the existing road.

The SCUT (Shadow Toll) Programme: General Description

The SCUT programme includes seven concessions in various regions of Portugal, all of them already contracted (see Table 1).

SCUTs are public works concession contracts for 30 years, between the state and private consortia. On the basis of these contracts, the private consortia are responsible for the operation of a public service (conception, construction, maintenance and financing of the sections of motorway described in the contract). At the end of the 30 years, the infrastructure is returned to the state without costs.

Table 1. SCUT Programme – General Data.

Concession	Length (km)			Status	Date of Contract
	New	O&M	Total		
Beira Interior	130	50	180	In service	13 September 1999
Algarve	36	92	128	In service	11 May 2000
Costa de Prata	63	38	108	In service	19 May 2000
Interior Norte	116	39	155	Under construction	30 December 2000
IP5-Beira Litoral/Alta	169	5	174	Under construction	28 April 2001
Norte Litoral	41	72	113	Under construction	17 September 2001
Grande Porto	49	15	64	Under construction	13 September 2002
Total	604	311	922		

In compensation, the state pays the concessionaire an annual amount computed on the basis of the number of vehicles that drove through the concession (utilisation) and on indicators of quality of service (performance). The structure of payments has been designed to respond to the objectives of the government and to the requisites of a project finance scheme.

These payments are based on a structure of three bands, where tariff levels (per vehicle kilometre) decline with the increase of traffic, with the purpose of reducing the commercial risk. Above a certain number of vehicles, total payment remains constant, thus limiting the upside for the concessionaire and the exposure for the state. In this model there is no guarantee of minimum traffic.

Performance bonuses (or fines) are foreseen in the contract, based on indicators of the performance of the concessionaire. Performance is evaluated based on accident rates and number of hours of lane closure for maintenance.

In addition to toll payments, the state is responsible for the approval of the projects, for supervision of the works and regulation of the concession and for some risks.[4]

The sources of financing for SCUT concessions were in general senior bank debt (85%), bonds (3%) and own equity (12%). The senior bank debt was contracted with the EIB and with commercial banks, which also issued the necessary guarantee for the EIB debt.[5] The high volumes of financing needed for these projects warranted that the debt structure was arranged by large banks (mostly Portuguese) with strong project finance experience (the 'arrangers'), subsequent syndication in the international market and with participation of large international banks.

The contractual structure associated to the SCUT model is complex, including dozens of contracts. The concession contract is the basic contractual instrument, to which several annex contracts are associated, establishing the relations and conditions among the involved parties.

Justification Presented by Government for the Introduction of the SCUT Concessions

The objectives described in the legal acts that launched the various motorway concessions are basically the increase of available financial, technical and human resources mobilised for the motorway programme and the anticipation and the conclusion of the planned network as defined in PRN 2000.

The main quoted reason for the launch of these partnerships with the private sector is the increase of investment volume and corresponding acceleration of the completion of the (national or regional) road plans.

In the studies preceding the launch of the SCUT programme (Hambros, Banco Efisa & Sousa Brito, 1997; Banco CISF, 1998) other motives were presented for recourse to the public–private partnership/project finance model:

- to ensure that projects were allocated quickly (accelerate conclusion of PRN 2000);
- to ensure that they were accounted for out in the balance of the state;
- to minimise annual payments by the state;
- to maximise the benefit–cost relationship for each of the projects;
- to develop in Portugal a competitive sector for operation and maintenance of motorways;
- to provide a more balanced split of risks;
- to provide access to know-how and technical innovation brought into the projects by the private partners;
- to improve quality of service to the users.

One of the major reasons to justify the SCUT regime was to accelerate the development of the Portuguese hinterland, which justified a larger effort from taxpayers.

The inexistence of free access road alternatives (of acceptable quality) was another of the reasons for non-application of tolls on these motorways, bearing in mind what had occurred in the west concession. In fact, although there is no legislation forcing the existence of free access alternatives, the state has always respected this principle,[6] even if it never worried to define in quantitative terms what this is supposed to imply.

The transaction costs involved in toll collection was another of the reasons for the choice of the SCUT regime. In fact, as of 1997, only traditional toll collection forms were available, which would lead, in some concessions, to excessive transaction costs in relation to the expected traffic volumes.[7]

Another (political) reason that may have weighted in the option for the SCUT regime in some concessions is the fact that they already had some sections under free access operation for a number of years. The introduction of tolls in those sections, or the utilisation of different regimes in adjacent sections, could have significant political risks (the memory of incidents on the 25 April Bridge in 1994 was still fresh), with obvious impacts on the evaluation of the project by the private partners and their lenders.

Characterisation of the Burden for the State in the SCUT Concessions

The burden for the state budget from the SCUT programme (payment of the shadow tolls to the concessionaires) is expected, in our forecast, to amount to about €550 million (at 2003 constant prices) for 15 years, from 2007 onwards, but as a percentage of GDP it will decline from 0.38% in 2007 to 0.20% in 2023. These two perceptions probably underlie the different political readings of this burden, and the associated policy orientations. For the forecast, GDP was projected on the basis of the Portuguese 2003 GDP of €130,033 million, growing at 1% in 2004 and at 2% from 2005 onwards. Inflation values used in the computations were the historical ones up to 2004 and 2% from 2005 onwards.

Socioeconomic Evaluation of the SCUT Programme

The set of the first six SCUT concessions[8] has already been subject to a published evaluation (Fernandes, Oliveira, & Lisboa Santos, 2002). For this purpose, the costs of the programme (i.e. the shadow tolls paid) were compared with the economic and social benefits generated (time savings, reduction of accidents, fiscal return on the construction phase). This work has shown that the benefits are larger than the costs (Fig. 1).

In the same work, the SCUT model was compared with the traditional contracting regime. For that, the costs and benefits associated with either model were estimated. With respect to the SCUT model, the shadow tolls and the benefits associated with the anticipation of opening to service, estimated to be 10 years, were considered. With respect to the traditional

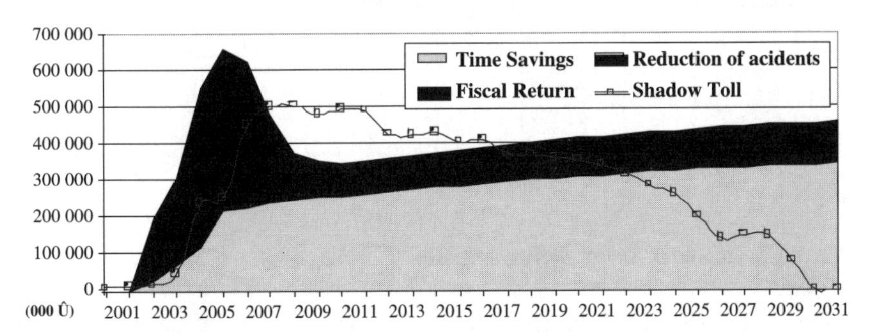

Fig. 1. Costs and Benefits of the Six First SCUT Concessions. (*Source:* Fernandes et al., 2002.)

contracting model the costs of construction and operation borne by the state during the concession period were considered.

The benefits and costs of both solutions (SCUT and traditional contracting) were discounted to the same base year, using several discount rates. This comparison allowed the authors conclude that the SCUT model was more efficient (see Fig. 2).

However, this comparison only analyses the efficiency and effectiveness of the SCUT programme, comparing it with two other alternatives: do nothing and build under traditional contracting. A comprehensive evaluation would further require considering the following issues:

- Other financing alternatives; for instance, application of real tolls.
- Individual project analysis – the fact that SCUT concessions as a whole lead to higher benefits than costs does not mean that some of them would not have it otherwise.
- Equity consideration – evaluation of the distribution of costs and benefits by these projects, by themselves and in comparison with what has occurred with the motorways with real tolls.

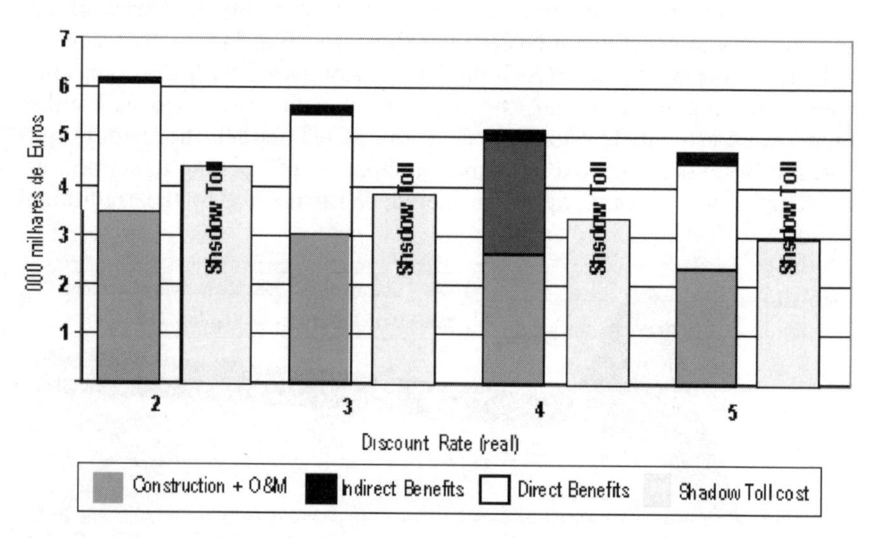

Fig. 2. Global Costs of the Six First SCUT Concessions vs. Equivalent if Done Under Traditional Contracting. (*Source:* Fernandes et al., 2002.)

Current Status of the National Motorway Network

The Portuguese motorway network (those in existence plus those planned) may be divided into four groups according to the financing/management model of each motorway (the percentage of total motorway length in each group is given in parentheses):

- SCUT motorways (29%);
- real toll motorways;
  - tolled sections (59%);
  - free access sections (9%);
- Motorways financed and operated directly by the public agency (currently Estradas de Portugal, E.P.E.) (3%).

Part of the real toll component has not yet been put under concession, waiting for a tender or for a decision on an earlier tender. The already built part of this component (25%) is currently managed by the public agency. But parts of the shadow toll components (45%) are still under construction.

EVOLUTION AFTER THE FALL OF THE GOVERNMENT RESPONSIBLE FOR THE LAUNCH OF SHADOW TOLL MOTORWAY CONCESSIONS

In April 2002, the newly installed Minister of Public Works announced that the shadow tolls concessions would have an unbearable weight on the government budget, and thus had to be converted into real toll concessions.[9] As it should have been expected, strong negative reactions followed from local politicians in the regions that were to have free access motorways and were now threatened with real tolls. In response, some other ministers and the prime minister himself stepped down, making public statements indicating that there would be some exceptions, sometimes speaking about the motorways serving the poorest regions and those which were already (partly) in operation and sometimes speaking about positive discrimination towards inhabitants and companies located in the regions serving (all) shadow toll motorways.

In spite of these statements, the government signed in September 2002 the contract for the last of the seven shadow toll concessions, the one of 'Grande Porto', invoking the urgency of the corresponding infrastructure.

In the background, some studies were initiated, dealing with the technical, economical contractual issues:

- As these motorways were conceived to operate without tolls, the average distance between their nodes was much smaller than on motorways with real tolls (typically between 3 and 5 km instead of 10 km). In combination with the relatively low traffic volumes, this made traditional toll collection methods, with a combination of electronic and manual collection, economically unfeasible: transaction costs would in several cases absorb large proportions of the total revenue. So, toll collection was to be possible with any of three automatic systems: DSRC[10] (Via Verde or similar), credit card and prepaid contactless smartcard (similar to one already in use at the Tagus bridges for regular travellers). These smartcards would be available at gas stations throughout the country, and especially advertised near these motorways. In parallel, vignette-based solutions (with different time spans and motorways validities) were also considered.
- The economic value of the shadow toll motorway concessions would be changed with the imposition of tolls, as some traffic would use alternative free-access roads, where they exist with acceptable quality. Given the low expected traffic volume in some of those motorways even under shadow tolls, and the need to provide decent alternatives, a careful economic analysis was necessary, not only about the necessary cost of the rebalancing indemnities, but also about the marginal value of installing toll gates at each node.
- There would be heavy contractual implications, related not only to the change of expected economic value of the concession, but also of its risk profile. The financing structure of the private side would thus have to be changed, even in those cases where the construction was already well under way or completed. The question of the charging mechanism was also raised, as the concessionaires insisted that they had to be in charge of that process in their concessions, and the government indicated a preference for a global toll-collection concession across all seven shadow toll motorways, expecting to reap benefits of economies of scale and strong international competition.
- Still with no decisions taken on the conversion process, the Minister of Public Works resigned in April 2003, invoking health reasons. His successor basically followed the same line, apparently with lower priority for ending the studies under way, and having to in the meantime inaugurate some sections of those shadow toll motorways.

At the end of May 2004, the Court of Accounts published its annual report of the activities of 2003, during which the Shadow Toll Motorway Programme was audited. The findings about the soundness of the launch of the programme were (Tribunal de Contas, 2004):

- The launch of the SCUT concession programme was not preceded by an evaluation of its economy, efficiency and effectiveness in comparison with the traditional model of public works contract.
- The government launched the tenders for the SCUT motorway concessions without previous approval of the environmental corridors.
- The high charges and risks undertaken with the SCUT motorway concessions put under risk the budget sustainability of these contracts.

The recommendations were of a rather general tone:

- Whenever project finance schemes are considered for financing of public projects, the state must demonstrate that such a solution generates value for money with regard to the traditional option of support on the state budget.
- The state must also strive to appropriately consider the environmental aspects in the earliest possible phase of the tenders for SCUT motorway concessions.
- The state must create mechanisms for budget control which allow the assessment of state financial engagements in SCUT motorway concessions.

The last recommendation seemed to vindicate the position of the government, and no changes in policy were announced.

However, some of these conclusions do not reflect the content of the audit report or reveal incomplete information about the SCUT process. In particular, the comparative evaluation of economic efficiency of the shadow toll option vs. traditional contracting, and the corresponding generation of value for money, had already been made by the Ministry of Public Works and its main conclusions had even been the subject of an international publication (Viegas & Fernandes, 1999). On the other hand, the court ignored that the government had stated that it wanted to make good use of the ingenuity of private companies for project cost reduction, through their engagement in the conception phase, and thus in the choice of the corridors whenever environmental approvals had not been obtained earlier. In fact, in only one of the sections of one of the SCUT motorways (interior north) were there changes to the corridor proposed by the winning consortium with significant financial impacts.

Insufficient consideration of mid- and long-term budget impacts is a general weakness of public finances in Portugal, as the country does not have a legal rule for taking into account future financial responsibilities unless they are formally recognised as public debt. But this is a general weakness, not specific of the SCUT programme.

In June 2004, following the resignation of prime minister José Manuel Barroso to become president of the European Commission, a new government was installed. The new Minister of Public Works announced a new impulse to this file, making it a political priority. The dimensions of positive discrimination were announced, with toll reductions only for residents and companies located in any of the municipalities adjacent to the SCUT motorways and with lower purchasing power per capita (only as long as it lasted, with periodic reviews over the life of the concessions) and only for their local trips, within a radius of 30 km of the (two adjacent) entry nodes requested by each of those residents or companies. Given the cost increases of the SCUT programme as denounced by the Court of Accounts, and the persistent difficulties of the successive governments in reaching annual budget deficits below 3% of GDP, as imposed by the EU Growth and Stability Pact, political commentators and the finance sector had already been convinced of the need for change of concession regime as one of the steps to relieve pressure on the government expenditure as early as 2005.

On 30 September 2004, a resolution of the Council of Ministers of the (right wing) government was approved, in which a very negative image of the financial implications of the shadow toll programme (launched by a left wing government) for the road administration was presented, and a few guidelines for the solution of the problem were issued. The main points in these guidelines were:

- The introduction of (real) road tolls in those motorways, under the 'user–payer' principle.
- The securitisation of credits over revenues of road assets.
- The creation of a fund for maintenance, upgrade and safety of roads out of concession contracts.
- A mandate was given to the Minister of Public Works to carry out these measures, and to open negotiations with the concessionaires of the SCUT motorways.

A calendar was defined, with full implementation of all these measures and approval of the new model for financing of all road concessions until the end of 2004, i.e. a delay of 3 months.

The commercial banks claim that at the time of signing the contracts the EIB did not consider the total spending with this programme as too onerous for the spending capacity of the government as a whole (Thomson Financial,

2004). It is true that the amount in question represents only less than 1% of the national budget, but when the Portuguese government is having such dire difficulties in controlling the budget deficit, suggesting that this money can be found within the road sector, which already generates a financial surplus, is not realistic: the financial surplus of the road sector is being used to cover financial deficits of other sectors, and since the overall deficit is still out of control, the overall sum is more negative than it should, and there is no slack available.

At the end of November 2004, the President of the Republic dissolved the parliament, based on the signs of instability of the government. The studies that were under way continued to be carried out, but no effective decisions were taken regarding the SCUT programme given the frail statute of the outgoing government and the well-known resistance in many areas of society to the introduction of real tolls in those motorways.

Legislative elections were held in February 2005 and the Socialist Party secured an absolute majority in parliament; the new government was installed in March. The new prime minister has declared that the SCUT programme was to be kept as defined in the existing contracts, i.e. with shadow tolls. No details have been given yet on the political options that will be made to mobilise the €500 million needed for the 2005 budget or the €800 million needed for the 2007 budget.

Just a few days after the instalment of government, the Governor of the Bank of Portugal declared that if the SCUT concessions were not to be changed, the revenue to cover the corresponding costs for the state had to be found somewhere else, the best solution possibly being through an increase of the purchase tax on motor vehicles or of the fuel duties. This has raised some negative comments related to the intromission of the governor in executive matters, as well as to the fact that citizens using real toll motorways would be paying twice. At the time of writing (mid-April 2005) no definite policy has been announced to come out of this gridlock. The revision on 20 March by the EU finance ministers of the Growth and Stability Pact, giving national governments more room to justify deficits above 3% of GDP, will certainly be used in whatever solution is found.

CONCLUSIONS

Looking back at 33 years (1972–2005) of motorway concessions in Portugal, two main periods can be identified: up to 1997, a single concessionaire (BRISA) of almost exclusively public ownership; after 1997, the privatisation of BRISA and the launch of several concessions to private companies through tender.

In the first period, the state had a contract with BRISA, but the statute of dominant and later exclusive shareholder led to a confusion of roles and certainly to a loss of efficiency in the application of public money.

In the second period, several strategic options have been tested, but the state has failed too often, mainly because of excessive hurry in making decisions and launching programmes and tenders: sometimes careful studies are missing at the time of opening the tender, sometimes the expropriation processes and environmental approvals are left for a later stage, with the risk of cost increases, sometimes the political or macroeconomic risks are not adequately assessed.

As mentioned above, in the case of the second Tagus crossing, the difficulties and risks of increasing the toll on the old bridge by 50% were not adequately considered. That increase was never made, and as a consequence the contract had to be renegotiated and the public part of the investment went from less than one-third of the total to roughly two-thirds. Given the high transaction costs of any public–private partnership, it is worth questioning whether, at such a share of investment costs between public and private, it still represents value for money. The tenders for concession of motorways under real tolls have been running rather smoothly, with only scattered incidents of judicial disputes between candidate consortia concerning the allocation of the concession. Road traffic has been growing steadily in Portugal, more or less in line with the GDP, and this allows for relatively low risks in these concessions. In the case of the shadow toll programme, what could have been an interesting idea if adopted in a more limited scale (building motorways in less densely occupied and less rich areas of the country, possibly under a somewhat degraded standard, as done in Spain with the Autovias), has been the cause of many disputes caused mainly by its scale of roughly one-third of the total motorway extension in the country, to be built in roughly 10 years, instead of 20–30 years if done under the usual contracting conditions.

The scale of the programme was over-ambitious and did not test both the need for this total coverage of the territory in 10 years, and the possibility to pay the corresponding concession rents without excessively affecting the yearly budgets. Moreover, the macroeconomic risks associated with a slump of the economic growth (as it has happened since 2002) were underestimated. On top of that, the excessive hurry in launching the tenders has implied cost overruns in expropriation and alignment redesign for environmental acceptance, thus aggravating the bill even further.

It is not known yet how the country will solve the problems created by those hasty decisions, and a lot has been learnt about mistakes that could be prevented. The main lessons can however already be drawn: legislation is needed to force computation and presentation of future financial responsibilities of the state as accompanying information of any governmental or parliamentary decision with implications at this level.

NOTES

1. Public–private partnerships embody a set of forms and models of lasting co-operation between the public and private sectors for the construction or operation of large infrastructure projects by the private sector or for the delivery of public services to the population by the private sector on behalf of the public sector (Grimsey & Lewis, 2002).
2. Although the variable term of the concession, from an initially estimated value of 23 years up to a maximum 33 years is a measure that allows a partial transfer of this risk.
3. This total does not consider the not yet built sections included in the concession contract with BRISA.
4. In the case of SCUT concessions the state assumes the risks for *force majeure,* expropriations, specific legislative modifications and unilateral concession modification. The risks associated to design, construction and operation are fully underwritten by the concessionaire. The environmental risk is jointly underwritten by the state and the concessionaire.
5. The Algarve SCUT concession was financed with equity, EIB debt and bond emission, without involvement of the commercial banks. The EIB loan and the bond emission were guarantee by the *monoline insurer* XL Capital Assurance, and constituted the first financing of an infrastructure in continental Europe based on a bond emission.
6. The PRN 2000 plan upgraded the classification to national roads of some 800 km of roads so that they would constitute a free access alternative to the itineraries where the government had decided to introduce tolls. This fact is an implicit recognition of the unacceptability of imposing tolls on a motorway without a free access alternative with the status of national road.
7. This power of this argument was then amplified with the decision to have a much higher density of nodes on these motorways, since there were no toll collection costs.
8. The SCUT concession of 'Grande Porto' was not included.
9. The government has also announced, at same time, its commitment to develop the High-Speed Railway Network, with a total investment volume much higher than the total for the SCUT programme.
10. Dedicated short-range communication.

REFERENCES

Banco CISF. (1998). *AE em regime de Portagem SCUT – Public session for presentation of the SCUT concession 'Interior Norte'*. JAE Headquarters, 13 March 13.

Fernandes, C., Oliveira, A., & Lisboa Santos, J. (2002). *O Programa de AE SCUT: Avaliação de custos e benefícios*. Lisboa: 2° Congresso Rodoviário Português.

Grimsey, D., & Lewis, M. K. (2002). Evaluating the risks of public private partnerships for infrastructure projects. *International Journal of Project Management, 20*, 107–118.

Hambros, S., Banco Efisa, & Sousa Brito, C. (1997). *Concessão de Estradas em Regime DBFO*. Lisboa: Secretaria de Estado das Obras Públicas/Junta Autónoma de Estradas, January 1997.

Lemos, T. D., Eaton, D., Betts, M., & Tadeu de Almeida, L. (2004). Risk management in the Lusoponte concession: a case study of the two bridges in Lisbon, Portugal. *International Journal of Project Management, 22*, 63–73.

Thomson Financial. (2004). SCUTled – European PPP review – Portugal's shadow toll pile-up, European PPP review SCUTs, *Project Finance International*, October 2004.

Tribunal de Contas. (2003). *Auditoria ao contrato de concessão BRISA*. Relatório n° 13/03, 2ª secção, April, Lisboa (available at www.tcontas.pt).

Tribunal de Contas. (2004). *Relatório de Actividades e Contas 2003*, Lisboa (available at www.tcontas.pt).

Viegas, J. M., & Fernandes, C., (1999). Private financing of road infrastructure – the Portuguese experience. *Transportation Research Record*, Paper no. 99-06787, August.

FINANCING AND REGULATING HIGHWAY CONSTRUCTION IN SCANDINAVIA – EXPERIENCES AND PERSPECTIVES

Svein Bråthen

INTRODUCTION

This chapter reports on the structure, costs and regulations of the highway system in the Scandinavian countries of Denmark, Norway and Sweden. Technical and financial information on the highway network is briefly presented. The presentation rests heavily on statistical information from the national transport ministries and public roads administration. The amount of data and number of maps and illustrations published on topics relevant to this chapter varies among the countries. The scope of the presentation will therefore vary among them.

The countries are quite similar when it comes to the regulatory regime for highway investments and operations. Therefore, their regimes will be described in a section that includes all three countries. Possible differences will be commented upon.

The national road authorities are responsible for planning road investments and operations. The parliaments are responsible for the final decisions on investments in the highway network. The construction and

Procurement and Financing of Motorways in Europe
Research in Transportation Economics, Volume 15, 175–186

ISSN: 0739-8859/doi:10.1016/S0739-8859(05)15014-8

maintenance activities are mainly contracted to private enterprises. Denmark has a few very large toll projects like the Great Belt and the Øresund Bridge that have opened great opportunities for regional development, but that also cause some concern with respect to risk and economic robustness. Norway has the largest number of toll-financed projects. They account for around 25% of the road investments. None of them are in the mega class as single projects, but the toll-financed investment packages in Oslo and other larger cities are of considerable magnitude. Sweden has not used toll financing much. The only projects are the Svinesund Bridge (under construction) and the Øresund Bridge; the first is in collaboration with Norway and the latter with Denmark. The toll fees are regulated by the authorities to maintain public control over potential monopoly power in most of the toll projects. Public–private partnerships (PPP) are not used much in any of these countries, but there are trial projects under way in Norway, and there is considerable interest in this financing regime in Denmark. Sweden appears to be more reluctant regarding the PPP arrangement even though such arrangements are currently under consideration.

First, a short description of the population structure is given. Second, the highway network, expenditure levels and regulatory frameworks for funding road infrastructure will be described. Third, the planning and financing regimes will be briefly discussed.

POPULATION

The size and shape of a highway network is closely related to how the population is distributed within a country. The Scandinavian countries have a quite different population structure.

Denmark is quite densely populated compared with the rest of Europe. The largest concentration of inhabitants is in the Zealand/Copenhagen area and in the middle of Jutland and Funen. The distances between these areas are rather short, particularly after the Great Belt link was finished in the late 1990s. The population density is around 12 times the density in Norway. In Sweden, the majority of the population is around Stockholm and Gothenburg and in the southern region. Sweden's regional policy in the 1960s entailed significant migration towards urban areas.

In Norway, the main part of the population is concentrated around the larger cities of Oslo, Bergen, Trondheim and Tromsø. There are scattered settlements especially along the coastline. The Norwegian regional policy has been based on the maintenance of a distributed population. Historically, it was necessary to keep up the regional development and settlement in the northern part of Norway for military reasons during the Cold War. In addition, Norway has a long coastline with industries based on natural resources like fisheries, oil and gas. Finally, there has been a political wish to maintain local communities based on agriculture. The latter is now under pressure both because the government wants to reduce the level of transfers to the agricultural industry and because globalisation and EC regulations make it more difficult to protect this domestic industry. Today, there is a strong tendency of migration towards the regional centres and the larger cities.

HIGHWAY AND MOTORWAY INVESTMENTS

Denmark

The Highway Network

The length of the main arterial highways is 1,660 km, while the regional road network has 10,000 km (secondary highways and county roads). The annual investments in the public roads network for the years 1976–2002 showed a declining trend from €350 million (at 2002 prices) in 1976 to €110 million in 2002. The network was extensively upgraded in the 1970s and the beginning of the 1980s. The highways' share of total public road investments is 35% on average, varying between 18% and 50% in single years. The investments shown above do not include the toll-financed projects described subsequently. For the years 2004–2007, around €80 million per annum will be used for maintaining the 1,660 km of national highways. This is a significant upgrade compared with the last few years. Total revenues from road user taxes (excluding tolls) were €4.5 billion in 2002, of which 32% (€1.4 billion) came from fuel taxes. The rest of the road user taxes consists of vehicle acquisition taxes and annual road licence taxes.

The Mega Toll-Financed Projects: The Great Links of the Great Belt, Øresund and Fehmarn Belt[1]

The construction of the Great Belt and Øresund links was funded by loans in Danish and international capital markets. The holding company for Great Belt A/S, A/S Øresund and Sund & Bælt Partner A/S is Sund & Bælt Holding A/S, which is responsible for the operations, maintenance and financial management of the subsidiaries. Incorporated on December 10, 1991, Sund & Bælt Holding A/S is entirely owned by the Danish state. For the Øresund link, the Danish and the Swedish states have a 50% shared ownership through Sund & Bælt Holding A/S (Denmark) and Vägverket and Banverket (Sweden).

The Danish state acts as a guarantor for the construction loans for the Great Belt project and the Øresund landworks. The loans for the coast–coast facility at Øresund, including the immersed tunnel and the Øresund bridge, are guaranteed jointly and severally by the Danish and Swedish states. These guarantees ensure a high credit rating and, therefore, favourable borrowing terms. On the other hand, the incentives for the private lenders to ensure adequate payback may be weakened because of these state guarantees.

The Great Belt Bridge

The Great Belt fixed link comprises two bridges and one tunnel. Construction work lasted for 10 years (1988–1998), with the rail section opening in 1997 and the motorway section in 1998. The project links Zealand and Copenhagen with Denmark's mainland. Construction costs for the Great Belt project totalled €2.9 billion at 1988 prices, corresponding to approximately €4.8 billion at 2002 prices. The road and rail links each account for roughly 50% of the overall costs. All construction costs, including interest, will be repaid with revenue collected from the users, i.e. motorists and the Danish National Railways Agency. In 2002, motorists paid toll fees of approximately €270 million, while the annual fee from the Danish National Railways Agency totalled €80 million.

At the end of 2002, Great Belt A/S' debt was around €5 billion, including interest. The link had a deficit with respect to user payments during 2002, and outstanding debt increased by around €200 million.

Assuming a stable growth in traffic, continuing low interest rates and an annual adjustment of toll fees at the road link, the entire debt is expected to be repaid by 2026, i.e. 28 years after the opening.

The Øresund Landworks

This project connects the Helsingborg and Malmö area in Sweden with Zealand and Copenhagen in Denmark. The construction period was from 1995 to 2000. The construction costs of the Øresund landworks total €0.7 billion at 1990 prices (corresponding to approximately DKK0.9 billion at 2002 prices). The landworks comprise the Øresund motorway and the Øresund railway to Kastrup airport.

The construction costs, including interest, will be repaid from an annual fee from the Danish National Railways Agency and from dividend paid by the Øresund Bridge Consortium. In 2002 the fee paid by the Danish National Railways Agency amounted to €11 million. The Øresund Bridge Consortium has yet to pay dividends.

At the end of 2002 A/S Øresund's debt stood at €1.2 billion, including interest. The entire debt is expected to be paid within 56–59 years.

The Øresund Bridge

The construction costs for the Øresund bridge totalled €2 billion at 1990 prices (corresponding to €2.5 billions at 2002 prices). The entire construction costs, including interest, will be paid by the users, i.e. motorists, the Danish National Railways Agency and the Swedish Banverket (the Swedish rail track operator). In 2002 motorists paid toll fees of €75 million, while the fees from rail operators totalled €53 million.

At the end of 2002 the Øresund Bridge Consortium's debt stood at €2.7 billion, including interest. The link had a deficit with respect to user payments during 2002, with an accumulated debt of around €200 million. The debt is expected to be repaid approximately 35 years after the inauguration of the bridge, i.e. in 2035.

Fehmarn Belt

The Fehmarn Belt division of Sund & Bælt Holding has acted as adviser to the Danish Ministry of Transport on issues relating to the construction of a fixed link across the Fehmarn Belt between Denmark and Germany. A cable-stayed bridge with a four-lane motorway and two rail tracks could be a viable technical solution although an immersed tunnel is an alternative. At a meeting in Berlin in June 2004, the Ministers of Transport of Denmark and Germany signed a joint declaration setting out a detailed framework for the further development of the Fehmarn link. The ministers agreed on a financing model comprising state-guaranteed loans. Further investigations will aim at clarifying whether the private element within a state-guaranteed model can be further strengthened, e.g. by transferring some of the financial risk to the private sector. The payback period and the robustness seem comparable with those of the mega projects presented previously.

Norway

The Highway Network

The length of the main arterial highways is 7,547 km (of which only 178 km are motorways with four lanes or more, mainly located around the bigger cities), while the length of the secondary highways is 19,417 km. The main highways are inter-regional trunk roads. Apart from the four-lane motorways, the Norwegian highway network is characterised by mainly two-lane roads of variable quality.

Private financing of public roads dates back to the 1930s. More than 100 toll projects have been carried out successfully. In general, there is one toll company for each project, organised as a non-profit limited company, with local municipalities and firms as shareholders. In a few cases, one company has the responsibility for several projects. In most cases, the funding of toll roads is split between public funds and funds provided by loans taken up by the toll company. The distribution between public funding and toll financing is determined after a conservative analysis of what the road users are able to pay, given a certain toll level and a maximum payback period of 15 years. If the toll road is a fixed fjord link, the maximum toll level is normally set at a 40% mark-up on the fares of a ferry service of comparable length. The incentives for local firms to engage themselves in toll companies are connected to the expected savings in transport costs because the road projects are supposed to be implemented earlier than would have been the case with public funding only. The concession for the toll collection is given by the Norwegian parliament for a limited period of time, normally 15 years. The role of the toll company is to provide the funds and collect tolls for paying off the mortgage. The Norwegian Public Roads Administration (NPRA) is the regulator of the tariffs, and NPRA is normally in charge of the construction activities. The debt is (with one exception) not guaranteed by the state. With one exception, the public road authorities are the owner of the tolled roads, and they also carry the operating and maintenance costs.

Up to the late 1980's, toll financing was mainly used to finance bridges and tunnels. Nearly all the projects were located in rural areas. Since then, the number of tolled roads and the amount of tolls paid have increased, and today some 25% of the annual funding of road construction comes from toll financing. Only a few of the projects have at times faced economic problems. One is in fact bankrupt, but most of them are running well. The problems have mainly been due to increases in interest rates and too optimistic traffic forecasts.

The tariff is regulated by a moderated cost-plus model where the NPRA sets limits to tariff increases. If for instance the interest rate should double, the costs may not automatically be passed on to the motorists. The toll company has to give a substantiated application for every toll increase.

After the investment is recovered, the tolls are abolished and the toll company is liquidated.

The increased use of toll financing is due to a number of factors, but traffic growth has nevertheless intensified the need to invest. In the largest cities of Oslo, Bergen and Trondheim, congestion led to the implementation of toll cordons in the years 1986 to 1991. Currently, the legitimacy of these cordons is for funding and not for traffic control even if there is a slight difference between peak and off-peak charges. Using the Oslo toll ring for controlling congestion is on the political agenda, but no final decision has been taken.

Today, there are 45 toll projects, and the number of projects is increasing. Norwegian motorists paid €400 million for road tolls in 2002 (some €165 per vehicle per year). The revenues from the toll cordons make up the main part of the funds. Smaller projects in the rural parts of the country, however, still make up the majority of the projects. The revenues are assigned to the specific project where the tolls are collected. In the coming years, NPRA wants to use road tolls as a strategic measure to upgrade the national trunk road system.

A substantial part of Norway's highway funds (25%) comes from road tolls. These projects count for 689 km of roads, and a further 60 km are under construction. The accumulated debt for the toll roads was €1.25 billion by the end of 2002. Some projects (mostly around cities) have automated toll collection systems, the others have other electronic or manual systems.

The annual public investment in highways has accounted for between €510 million and €560 million during the years 1999–2005, and an annual average of around €550 million is planned for the next 10 years. Annual maintenance and operation costs are approximately 30% above the annual investments. Total revenues from road user taxes (excluding tolls) were €4.7 billion in 2003, whereof 38% (€1.8 billion) came from fuel taxes. The rest of the road user taxes consists of vehicle acquisition taxes and annual road licence taxes.

Sweden

The Highway Network

The Swedish road network comprises 138,000 km of public roads, of which 98,000 km are national roads. The motorways account for 4,800 km, while national highways account for 10,200 km.The rest (83,000 km) are county highways. The motorways are four-lane roads, while the highways are two-lane roads.

Sweden does not have toll-financed roads except for the cross-border engagement with Denmark and Norway (the Øresund Bridge and the Svinesund Bridge, respectively).

The exact total revenues from road user taxes could not be obtained at the time of writing; the tax structure is however similar to those in Denmark and Norway.

THE FRAMEWORK FOR ROAD PLANNING AND FUNDING

General Aspects/Comments

The organisation of the road-planning regime is quite similar in the three countries. The public sector, represented by the national road administrations (NRAs), is responsible for the planning, investments and operation of the national highway system. We will not examine the planning and decision-making process in detail in this chapter, but will pay attention to a few elements.

The NRA makes investment plans for the highway system. In Norway and Sweden, the regional level (the county) submits the priority of projects in the secondary system to the NRA through a political process. The NRA may overrule the regional priority lists, but as a rule they are followed and included in the NRA's road investment plan for each county. The primary highway system consists of the interregional trunk roads and is the NRA's responsibility, but the counties and municipalities are consulted when issues like land use and environmental issues are affected. In Denmark, the NRA is responsible for the interregional highway network (1,660 km or 2% of the public road network, which accounts however for 30% of the road traffic). The rest of the network is left to the counties and the municipalities. However, the monitoring of the road system, collection of traffic data, R&D for road construction methods, regulatory issues and international cooperation are the NRA's responsibility.

As shown in the foregoing, there are different practices when it comes to the private sector's role in toll financing and PPPs in the Nordic countries. Denmark has used full toll financing, private consortia and user payments with state guarantees in large projects. Sweden's only toll-financed projects are cross-border projects with Denmark and Norway.

Norway has 46 toll-financed projects in operation, and the number is expected to increase. With a few exceptions, the state has not provided any guarantees for the loans, but a system for risk management is established that provides the incentive to give careful estimates on the project's payback ability. The model for conditional reimbursement has been used as a vehicle for risk sharing between private agents and the authorities. These agreements have been framed by the Ministry of Transport and Communications as a way of providing local guarantees within adequate financial limits. The basis for such agreements is that both the project and the model for private financing are approved by the local public authorities. The main components in the model are the following. (1) The share of private financing in the projects is set with the help of conservative estimates of the traffic volumes with respect to induced traffic and traffic growth. (2) The share of public funding from the Government may be advanced by the toll company, with later refunding by the Government. (3) If the toll revenues exceed the careful estimates from (1), then the refunding in (2) is wholly or partly dropped. These public funds are spent on other highly preferred projects within the same region instead.

The absence of state guarantees combined with the possibility of using these funds at the regional level strengthens the incentives to make conservative risk estimates. However, if the toll revenues turn out to be insufficient, no additional governmental grants will be provided.

The toll level and discount system has to be approved by the NPRA for each project. If the traffic develops more positively than expected, the toll levels may be kept constant and/or the toll collection period may be shortened. NPRA has commissioned the development of an electronic toll collection system (AutoPASS). This system is extensively used in new projects and in those of the older ones that have a potential for a more cost-efficient toll collection. Converting the present toll cordons into congestion pricing schemes in the larger cities is a matter of consideration, even if it is politically controversial. The Norwegian parliament has passed an amendment to the Road Traffic Act that allows congestion pricing to be implemented.

PPPs are not commonly used in the highway network in Scandinavia. Denmark has one project under way, a highway project under the jurisdiction of South Jutland county. Norway has two pilot projects under construction, and a third will soon be launched, under the jurisdiction of NPRA. Sweden has alternative financing models like toll financing and

PPPs under consideration for the highway system as well as for other parts of the transport sector. Up to now, all projects in Sweden apart from the cross-border bridges are financed by state funds.

A Few Experiences from Norwegian Road Tolling

Even though a large number of Norwegian roads have been fully or partly financed with road tolls during the last decades, the way toll road projects have been organized and implemented has been subjected to criticism. The funding and toll collection is in many cases performed by private "local enthusiast" companies that operate on a non-profit basis. There is a project-specific relationship between a toll road project and a toll company, and the company is responsible for the economy in that particular project. The company is liquidated when the toll collection period ends. Even if this prevents cross-subsidisation, it is not necessarily the most efficient way of collecting tolls. Higher revenues than expected (from an increase in traffic and/or efficient management) will inevitably cause an earlier stop in the toll collection and a discontinuation of activities for the company. Conditions like these do not give the right incentives for more efficient management.

In a 1999 report the Office of the Auditor General of Norway pointed out several weaknesses and incentive problems connected to financial management and the organization of the toll companies as well as the apportionment of liability between the toll road companies and the authorities. Some of this criticism was repeated in a report in 2004. NPRA has advocated the need for changes in the current organisational framework, where NPRA should play a more active role as a coordinator and take more extensive responsibility for the management of the toll collection systems through a public toll road company.

Differences with respect to toll collection efficiency exist. The operating charges' part of the revenue varies from 5% to more than 35%, and the collection cost per trip ranges from €0.13 to 3.75. These observations illustrate that there may be undesirable differences even if some of the differences have to do with scale effects.

PPPs in Norway – Opportunities and Experiences

The state is the owner of the PPP roads, but the responsibility for construction and maintenance is contracted to a private consortium. In the trial projects, private consortia build, maintain and operate the road system. For the 45 tolled projects already in place, NPRA builds and operates the roads. The only role of the toll companies in these cases has been to provide funding and be responsible for the toll collection. A combination of public funding and road tolls is the chosen funding model in the PPP projects. The share of public funding is set to avoid too high tolls resulting in inefficiency from traffic deterrence. A road toll is used as the user payment vehicle (around €2 for cars, around €4 for HGVs, with an up to 50% discount, depending on the level of the pre-payment). The value of the contracts for the two projects under construction is between €145 and €175 million. The PPP consortium is reimbursed annually, and no payments are made until the new road is opened. Criteria for reimbursement are established, where incentives are given for providing a road of adequate standards and reducing the number of accidents. There is also a malus possibility if the standards are not met. None of the PPP projects are finished yet, so the efficiency of these contracts remains to be seen.

There are a number of circumstances that may affect the efficiency of PPP contracts. First, the authorities bind themselves to a monopolist for the contract period (25 years), and the contracts should be designed carefully to avoid strategic behaviour. Another uncertainty is connected to the stability of the private contractors under long-term contracts. Bankruptcy, M&A and flagging-out are factors that may cause regulatory problems. A third element is connected to the degree of flexibility with respect to endogenous conditions: projects in adjacent areas may affect the choice of the route or mode and hence the traffic volume in the PPP project, causing prospective needs for renegotiations of the contract.

As for toll financing, PPPs will contribute to relieving public budgets and at the same time get important projects implemented. This is the main reason why Norway is trying to use PPPs. Another main reason is connected to an expected increase in efficiency in the construction process and also possibly shorter construction periods. On the other hand, extensive use of PPP-financing will commit future parliaments economically and hence reduce future degrees of freedom with respect to policy design. This commitment occurs because the projects are partly publicly funded. In Norway, the three pilot PPP projects are expected to bind about 10% of the road investment budget for the years 2006–2012. An increase in the number of PPP projects today can therefore add power to the present parliament at the expense of future parliaments. NPRA does not want to start more PPP projects before there has been a thorough evaluation of these pilot projects. Indications from UK experiences (five projects) varied from a gain of around 30% to a loss of around 20% (Institute for Public Policy Research; referred in Aas, 2003) as compared with traditional contracts. The conclusion was that the gains were uncertain and that it was too early to conclude whether PPP will give added value as compared with traditional projects.

NOTES

1. Fact source: www.sundogbaelt.dk. The considerations are the responsibility of the author.

REFERENCES

Aas, H. (2003). Storbritannia: Usikre innsparinger ved offentlig privat samarbeid. *Samferdsel, 4*, 20–21.

IS A MIXED FUNDING MODEL FOR THE HIGHWAY NETWORK SUSTAINABLE OVER TIME? THE SPANISH CASE

Germà Bel and Xavier Fageda

INTRODUCTION

Toll motorways in Spain are heavily concentrated in two transport corridors, the Mediterranean Coast and the Ebro River valley. High-capacity road services are tolled in some territories and are free in others, while quality is similar everywhere. User tolls were used to finance the first expansion of the motorway network in the 1960s and early 1970s. The second wave of the network expansion took place in the late 1980s and early 1990s and depended on the public budget for funds in this period. Since the late 1990s, public financing has continued to be the main funding source for new motorways, although some have been financed through user tolls. In essence, the policy of recent years has combined, expanding the network of free major roads while continuing to allow private firms to construct toll motorways.

Because of this irregular pattern of funding, the motorway network in Spain is quite singular among the most populated European countries, with mixed funding sources used to finance the building of new motorways and

Procurement and Financing of Motorways in Europe
Research in Transportation Economics, Volume 15, 187–203

ISSN: 0739-8859/doi:10.1016/S0739-8859(05)15015-X

the maintenance of old and new motorways. Rounding up, 80% of motorways have been built and are maintained with public funds, while 20% have been built and are maintained with user tolls. With little variation in the level of road services, tolls are charged in some territories but not in others. This results in an unequal treatment of the road user and damage to the competitive status of the firms located in the toll territories.

Highway policies should be more rational, but it is not clear how progress can be made toward the functional and financial homogenization of the motorway network in Spain. The main goal of this chapter is to examine this issue. First, we briefly review the history of toll motorways in Spain. Second, we analyze the different models of highway financing implemented since the 1960s. Then, we characterize the structure and regulation of the motorways business sector. Based on the previous analysis, we discuss different policies that could be applied to produce a more homogenous system of highway finance and management in Spain.

HISTORY OF TOLL MOTORWAYS IN SPAIN: PROMISES AND RESULTS

In the 1960s the Spanish economy was involved in a structural transformation, in accord with the 1959 Stabilization Plan. The economy was growing fast and transportation infrastructures were an increasingly narrow bottleneck for productive activities. The World Bank (1962) report on economic development in Spain recommended an effort to repair and maintain the existing road network. The World Bank also suggested the building of a new motorway along the Mediterranean coast, from the French border to Murcia. This road would serve important industrial and agricultural areas as well as some of the most important tourist destinations of the country. It would cross those territories with the greatest and most quickly increasing traffic in Spain.

In 1967, the Government planned for 3,160 km of toll motorways in the Program of Spanish National Motorways (PANE). Up to 1972 the sections franchised to private firms comprised La Junquera (French border)–Barcelona–Tarragona, Mongat–Mataró, Bilbao–Behovia, Villalba–Villacastín–Adanero, Seville–Cadiz, and Salou–Valencia–Alicante. The possibility of having motorways (even if tolled) raised great expectations, and political and institutional pressures to acquire such roads emerged all over the country. The PANE update of 1972, the Advance of the National

Table 1. Toll Motorways Concessions until 1975.

Concessionaire	Section	Period	Term (years)
Haches	La Junquera–Barcelona	February 6, 1967	37
	Montgat–Mataró	February 6, 1967	37
	Barcelona–Tarragona	January 29, 1968	37
	Montmeló–Papiol	1974	—
Iberpistas	Villalba–Villacastín	January 29, 1968	50
	Villacastín–Adanero	September 30, 1972	50
Europistas	Bilbao–Behovia	March 23, 1968	35
Bética de Autopistas	Sevilla–Cádiz	July 30, 1969	24
Marenostrum	Salou–Valencia	September 8, 1971	27
(Aumar)	Valencia–Alicante	December 22, 1972	27
Audenasa	Tudela–Irurzun	June 8, 1973	41
Audasa	Ferrol–La Coruña–Santiago–Pontevedra–Vigo–Tuy	July 18, 1973	39
Acasa	Zaragoza–El Vendrell	July 25, 1973	25
Vasco-Aragonesa	Bilbao–Zaragoza	November 10, 1973	22
Eurovías	Burgos–Malzaga	June 26, 1974	20
Aucalsa	Campomanes–León	October 17, 1975	46

Note: Tudela–Izurzun depends partially on the Navarre local government. Aumar took over Sevilla–Cádiz and Acesa acquired Zaragoza–El Vendrell. In 1976 Bilbao–Santander was provisionally franchised, but the final franchise was not undertaken.
Source: Bel (1999).

Plan of Motorways, included 6,340 km of toll motorways. Promises were high, but results did not meet expectations. Table 1 shows the concessions franchised up to the end of 1975. They add up a total of 2,042 km.

However, operational kilometers of toll motorways were slow to open. In fact, no more than 1,807 km of toll motorways were operating by 1985, along with 1,363 km of free motorways. To sum up, by the late 1960s and early 1970s, there was a general desire for motorways, and national government planning attempted to satisfy almost every single demand. We are left to ask why the reality was finally so modest if the proposals were so ambitious?

MOTORWAYS IN SOUTHERN EUROPE: TOLLS VERSUS GENERAL TAXATION?

By the middle of the 20th century, motorways financed in the public budget were not the general pattern in Mediterranean Europe. The most populated countries in Southern Europe, Italy and France, chose to finance motorways through user tolls. Even so, the networks were publicly owned and managed. In the 1960s, Spain also chose to finance motorways through tolls. Why did the southern countries choose tolls instead of public budget financing?

Budgetary financing of infrastructure has two inter-related requirements: a) the political will to levy general taxes and b) the availability of a modern and efficient tax system, so that public revenues are sufficient to finance such policies. South European countries have usually been less willing to use the general tax system than countries from Northern and Central Europe. Furthermore, the tax systems in the Mediterranean countries were the least efficient of the Western European countries in the 1960s and 1970s. Indeed, tolls were used to finance motorways because public budget constraints and a lack of political willingness to increase tax revenues made tolls the only option.

In Spain, the shortcomings of the tax system and the lack of willingness to upgrade it made it difficult to use the public budget to finance motorways. The PANE of 1967 already opted for financing through user tolls. Even in this case, a model of public management could have been applied, as in France or Italy. Spanish private firms used government loan warranties to obtain funds abroad, showing clearly that the state had the same or better access to external funds as private firms. Nonetheless, the Spanish government made a choice that was exceptional in that period: to award the building and operating of motorways to the private sector.[1]

In fact, this concession did not insulate the public budget from the risks and costs of financing motorways. Numerous financial, fiscal, and commercial conditions transferred almost every risk from the private firms to the state. The insurance for the exchange rate in external debt has been especially damaging for the public budget. Indeed, the Spanish history of motorways is a stark demonstration of the constraints and costs to the Treasury that can emerge from a system of private toll motorways. The government's long-term commitments with private firms led to huge payments. These commitments induced inefficient economic decisions, whose costs were, and are still, borne by the Treasury.[2]

CHANGING MODELS SINCE THE 1980s: FROM USER TOLLS TO PUBLIC BUDGET

By the mid-1970s, some of the shortcomings of the toll model had already emerged. The 1974 report about national toll motorways pointed out the reduction in the rate of traffic growth, the increase in the price of external debt, and the high building cost of the Spanish toll motorway network. Gómez-Ibáñez and Meyer (1993, p. 131) report: "In several cases, construction costs had been four or five times original projections, while initial traffic volumes were as little as one-third of those expected."[3] Once the economic crisis arose, what private firm would be willing to invest in motorway sections when demand forecasts were lower than they had been for the sections first franchised? Not surprisingly, the concessions suddenly stopped. The 2,042 km franchised up to 1975 (not all of them in effective operation) did not rise until 1987, when a new motorway was franchised by the regional government of Catalonia. The territorial distribution of toll motorways is a consequence of the private model of finance and management. In general, sections with the highest expectations of profitability were the first to be franchised, and the concession process broke down when the crisis arose. No one had an interest in "wasting" money to invest in corridors with low expectations of profitability.

This breakdown of concessions is typical of private systems of tolls. Each section is franchised on a separate basis, and its profits or losses are individually considered. In contrast, in France and Italy, public management allowed a network, rather than individual, approach to concessions. In Italy, the profits of some routes were used to expand the network into less profitable sections. This explains why the economic crisis of the 1970s slowed down, but did not completely break down, the expansion of the Italian motorway network.

The Socialist Party (PSOE) won the 1982 election, and a state-owned firm, Enausa (ENA), was created in 1984 to take over three private concessions that had gone bankrupt and were unable to develop their franchised roads: Audasa, Audenasa, and Aucalsa. As a rule, the socialist government chose a model of public financing of motorways in the 1984–91 Roads General Plan. With this choice made, the model of motorways financing moved toward the usual model in Northern and Central Europe and the Anglo-Saxon countries. Three reasons could explain this change:

1. *Fiscal feasibility*: Creating and enforcing the income tax in 1977 had been a huge step toward overcoming the backwardness of the Spanish tax system. With the available modern fiscal tools, the new government chose to put fiscal pressure on the economy closer to the European Community (EC) average. This made possible the public financing of motorways, among other programs.

2. *Fast delivery of motorways*: Private toll motorways had delivered modest results. Between 1970 and 1985, some 1,700 km of toll motorways were built, and the total network (including free motorways) consisted of 3,170 km.With the new model, expansion of the network has been much more rapid. In just 7 years, the network multiplied by 2.2, due to the addition of about 3,600 km of free motorways between 1986 and 1992. During the 1990s and the beginning of the new century, the supply of free, high-capacity roads has been growing rapidly.
3. *Availability of EC funds*: Within the context of the public financing model, there was another relevant factor: the four regional areas in which more sections of motorways were built between the mid-1980s and the mid-1990s are Castile-La Mancha, Andalusia, Castile-Leon, and Valencia. All were regions included in Objective 1 of the EC. This allowed the government to obtain high levels of co-financing from the EC through the Regional Structural Funds.

Alongside the general trend toward budget financing, some specific new policies in favor of tolls have been implemented since the early 1990s. We can outline (1) the re-negotiation agreements for extending the period of the concessions and especially (2) the 1997 Program of Toll Motorways, drawn up by the government of the Popular Party (PP, conservative), in power after the 1996 national election and until mid-2004. Agreements between government and private firms to extend concessions were widely used during the 1990s. Indeed, at the end of 1996 a national law allowed concessions to be extended for up to 75 years, which has promoted the use of this type of agreement.

Each agreement was implemented through direct re-negotiation between the government and the concessionaire, since the EU rules regarding competitive procedures to extend concessions were not binding at that time. The firms were not required to pay any fee to the state for having the concession extended. Usually, concessions were extended either to compensate for reducing toll prices or in return for the concessionaire's willingness to undertake unsound investments in other motorways. Indeed, the extension agreements resulted in huge profits for the private firms; they had their businesses extended on very favorable conditions. The amount of investment agreed upon and some reductions in tolls do not justify the large increases in the term of the concessions. Indeed, toll reductions stimulate traffic increases.[4] Since the marginal operating cost of a highway is very low, increasing traffic partially compensates for any toll reduction. But this was forgotten when negotiating the agreements.

Finally, within the context of the new policies in favor of developing tollways, we must mention the 1997 Program of Toll Motorways. Even if this program was a real deviation from the former policy of (almost) no new toll motorways, it does not signal that the conservative government has dramatically changed the model of public budget financing. In fact, the government acknowledged that many of the new toll motorways franchised to the private sector needed huge subsidies from the treasury because of low traffic (current and future).

Second, and more importantly, the 1997 Program of Toll Motorways did not imply an end to the expansion of the free motorways network. In fact, the percentage of free motorways increases from 76% at the end of 1996 to 79% at the end of 2003. If we focus only on the sections of motorways that became operative between 1997 and 2003, 85.9% were free motorways, whereas only 14.1% were toll. Therefore, the overall percentage of toll motorways has decreased. Table 2 displays the evolution of the Spanish motorway network. Fig. 1 shows the territorial distribution of national toll motorways and national free motorways in 2002.

The socialist party (PSOE) won the March 2004 national election. Although no policy against the currently operating toll motorways is expected, there will likely be a downsizing of proposals in the 1997 National Toll Motorways Program that have not yet been implemented.

CURRENT STRUCTURE AND REGULATION OF THE TOLL MOTORWAY BUSINESS IN SPAIN

Motorway Sector Structure

Toll motorway franchises in Spain amounted to almost 2,900 km at the end of 2004, although some of the concessions in Table 3 are not in effective operation yet. Around half of these kilometers belong to Abertis, the largest private Spanish firm in this sector. Abertis holds 1,240 of the franchised kilometers (43% of the total). Itinere Infraestructuras, owned by the holding Sacyr-Vallehermoso, is the second largest group by length of concessions. Itinere bought in June 2003 the four concessions of the public firm ENA (Audasa, Audenasa, Aucalsa, and Autoestradas de Galicia). Itinere concessions amount to 467 km (16% of the total). In addition, Abertis and Sacyr-Vallehermoso jointly control Avasa (294 km, 10% of the total). Finally, Europistas is a third important group because it is a significant shareholder of concessionaires such as Autosol and Autopistas Madrid Sur.

Table 2. Evolution of the Spanish Motorway Network Length (km).

Year	Total Motorways	Toll Motorways	% Toll/Total	Free Motorways	% Free/Total
1970	203	82	40	121	60
1975	888	619	70	269	30
1980	1,933	1,530	79	403	21
1985	3,170	1,807	57	1,363	43
1990	5,126	1,898	37	3,228	67
1991	5,801	1,957	34	3,844	66
1992	6,988	1,991	28	4,997	72
1993	7,404	1,991	27	5,413	73
1994	7,736	2,011	26	5,725	74
1995	8,133	2,023	25	6,110	75
1996	8,503	2,023	24	6,480	76
1997	9,063	2,063	23	7,000	77
1998	9,649	2,072	21	7,577	79
1999	10,306	2,239	22	8,067	78
2000	10,480	2,239	21	8,241	79
2001	11,152	2,277	20	8,875	80
2002	11,406	2,386	21	9,020	79
2003	12,009	2,517	21	9,492	79

Note: Since 1985, free motorways include roads of four lanes that were not previously labeled as motorways. Hence, it should not be implied that the extension of free motorways was high in the early 1980s. Actually, there was no real increase.
Source: Ministerio de Fomento (2004).

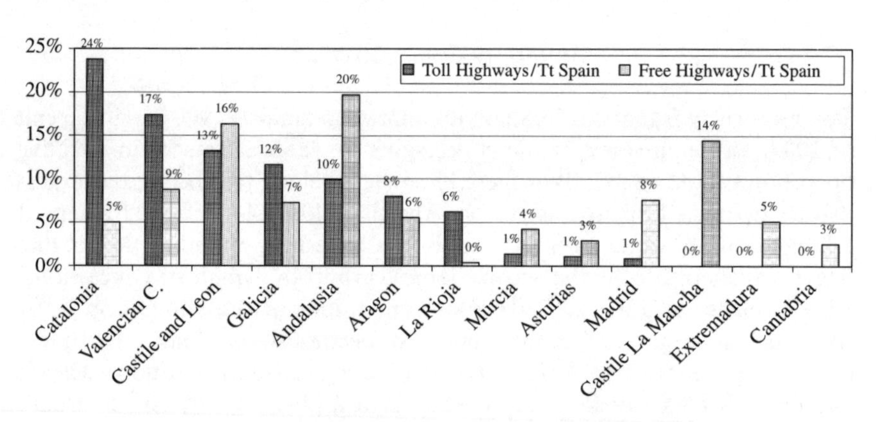

Fig. 1. Territorial Distribution of Free and Toll Highways.

Table 3. Toll Motorway Concessions in Spain in 2004.

Build–Operate–Transfer (BOT)				
Concessionaire	Section	km	Price per km (€)[a]	End of Concession
Acesa (Ab1)	Barcelona–La Jonquera	150	0.07	2021
	Barcelona–Montmeló	14	0.08	2021
	Montgat–Palafolls	49	0.07	2021
	Montmeló–Papiol	27	0.00	2021
	Barcelona–Tarragona	100	0.06	2021
	Zaragoza–Mediterráneo	215	0.09	2021
Aumar (Ab2)	Tarragona–Valencia	225	0.08	2018
	Valencia–Alicante	149	0.07	2018
	Sevilla–Cádiz	94	0.06	2018
Aucat (Ab3)	Castelldefels–El Vendrell	58	0.08	2039
Castellana (Ab4)	Avila–Villacastín	23	0.03[b]/0.05[c]/0.07[d]	2031
	Segovia–San Rafael	70	0.04[b]/0.08[c]/0.11[d]	2031
A-6 (Ab5)	Villalba–Adanera	28	0.11	2018
Aulesa (Ab6)	León–Astorga	38	0.09	2057
Avasa (Ab&SV)	Bilbao–Zaragoza	294	0.08	2026
Audasa (SV 1)	Ferrol–Portuguese border	219	0.06	2048
Audenasa (SV 2)	Tudela–Izurzun	113	0.08	2028
Aucalsa (SV 3)	León–Campomanes	78	0.12	2050
Autoestradas de Galicia (SV 4)	A Coruña–Carballo/Puxeiros–Baiona	61	0.05	2045
AutoSol	Málaga–Estepona (a)/Estepota–Guadiaro (b)	105	0.06 (a)/0.07 (b)	2049/2057
Accesos Madrid	Madrid–Arganda(a)/Madrid–Navalcarnero (b)	93	0.05/0.06 (a)/0.06/0.08 (b)	2049
Europistas	Burgos–Armiñón	84	0.10	2017
Autopistas Madrid Sur	Madrid–Ocaña	88	0.03/0.07	2069
Autopista Madrid-Levante	Ocaña-La Roda	118	n.o.[e]	n.o.
Enrasa	Madrid–Guadalajara	81	0.05/0.06	2028
Ausur	Alicante–Cartagena	77	0.02/0.03	2048
Autopistas Madrid-Toledo	Madrid–Toledo	60	n.o.	n.o.
Central Gallega	Santiago–Alto Santo Domingo	57	0.08	2049
Autema	Sant Cugat–Manresa	43	0.06	2039
Tunnel of Cadí	Tunnel of Cadí	30	0.30	2023
Tabasa	Tunnel of Vallvidrera	17	0.12[c]/0.13[d]	2037
Autopista eje aeropuerto	Eje aeropuerto	8	0/0.15	2030
Tunnel of Artxanda	Tunnel of Artxanda	5	0.19[c]/0.27[d]	2052
Tunnel of Sóller	Tunnel of Sóller	3	1.18	2016
Total		2,758		

Table 3. *(Continued)*

Build–Operate–Transfer (BOT)				
Concessionaire	Section	km	Price per km (€)[a]	End of Concession
Management contracts				
Outsourcing Manager	Section	km	Price per km (€)	End of Contract
Europistas	Bilbao–Ermua	45	0.06	2013
Bidelan	Ermua–Behovia	70	0.07	2013
Total		115		

Note: The end of the concession can be conditioned upon the beginning of the motorway operation. This can introduce small variations in the final data. Because of this, the ending date for the concessions Madrid–Levante and Madrid–Toledo is not yet available.
Source: Own elaboration based on information in the website of the Spanish Association of Tunnels and Motorways (Aseta) and the websites of the concessionaries. SDC (2003) for Audasa, Audenasa, Aucalsa, and Autoestradas de Galicia.
[a]Prices refer to light vehicles.
[b]Off-peak time price.
[c]Regular price.
[d]Peak time price.
[e]n.o., Not in operation in 2004.

Table 4. Profits from Regular Activities.

1990	1995	2000	2001	2002
130.77	337.43	538.89	566.22	644.17

Note: Spanish concessionaires (million €). Concessionaires of regional motorways in Catalonia and Galicia are not included.
Source: Ministerio de Fomento (2003).

Since the early 1990s, a strong record of profitability has characterized the motorway business in Spain. Table 4 shows firms' profits in the sector. They range from €130 million in 1990 to more than €600 million in 2002. In addition, it must be said that the Spanish stock market has supported the development of the most important firms in the motorways sector, such as Acesa (now merged with Aumar/Aurea into Abertis) and Europistas. Acesa began to trade in the Madrid stock exchange in 1987 and Europistas in 1988.[5] These firms deserve more attention because of their prominent role in the Spanish motorway business.

Since the late 1990s, Acesa and Aumar (later Aurea) began to expand, developing their presence in new markets in Spain, Europe, and Latin America. In April 2003, Acesa and Aurea merged, bringing forth Abertis. Currently, the Abertis motorway network covers a high proportion of the toll roads in Spain, with a turnover representing between 70 and 80% of the total business in the sector. The group is composed of Acesa, Aumar, A-6, Aucat, Castellana, Aulesa (and Avasa jointly with Sacyr-Vallehermoso). Furthermore, Abertis holds stakes in Autema, Accesos Madrid, Henarsa, Central Gallega, and Cadi Tunnel. In Europe, Abertis has undertaken strategic alliances with major private operators, such as Autostrade in Italy and Brisa in Portugal, with a capital share of 8 and 10%, respectively. In the United Kingdom, Abertis holds a 25% share in R.M.G. And Abertis is present in Latin America with stakes in motorway operators in Chile (Elqui), Colombia (Coviandes), Argentina (Ausol), and Puerto Rico. Abertis shares (like those of Acesa and Aumar before) have usually been among those with the highest relative yield and most stable growth in the Spanish stock exchange.

As we have mentioned above, Itinere Infraestructuras is part of the holding company Sacyr-Vallehermoso. The acquisition of ENA in 2003 has substantially increased the involvement of such holding companies in the motorways business. Currently, Itinere is composed of the concessionaires formerly owned by ENA and is a shareholder of Henarsa, Autopistas Madrid Sur, Accesos de Madrid, and Central Gallega. Itinere also holds stakes in concessionaires in Portugal (Lusoponte, Autoestradas del Atlántico, and Via Litoral), Brazil (Triangulo do sol and Via Norte), and Chile (Elqui and Los Lagos).

Europistas was created in 1968 to develop the Bilbao–Behobia franchise, one of the first toll motorways to be in effective operation in Spain. In 1974 Europistas was part of the consortium that obtained the Burgos–Armiñón–Málzaga franchise, which has been fully operational since 1984. The concessionaire of this motorway was Eurovías, in which Europistas held 35.1% of the capital. Europistas took over Eurovías in 2002. In addition, Europistas holds stakes in Autopistas del Sol, Autopistas Madrid Sur, and Autopista Madrid-Levante and manages Artxanda Tunnels.

Finally, a new kind of management contract was developed in 2003. Since the toll motorway concession Bilbao–Behovia ended in June 2003, the local governments of Bizkaia and Guipuzkoa are in charge of the motorway sections in their territories. Each government created a public entity for this purpose (Interbiak and Bidegi). In turn, these entities called for tenders to maintain and operate their sections. Autopistas de Bizkaia (whose main shareholder is Europistas) won the tender in the Bilbao–Ermua section, and Bidelan won it in the Ermua–Behovia section. In both cases, the management contract will be in force for 10 years. The revenues of the outsourcing managers are composed of two components: one fixed and the other varying with traffic flows. Direct tolls are still charged. Given the prices, the local governments will enjoy huge net revenues from the tolls paid by users.

Motorway Regulation: Institutions and Rules

There is no specific and autonomous regulatory body for toll motorways in Spain. The Spanish Ministry of Fomento (responsible for public works and transportation) is in charge of specific sectoral regulation and supervision on national toll motorways. Monitoring is organized in the same way at the regional level.

The initial price of tolls has depended on the initial conditions in the concession and, thus, it has been set on an individual basis. In addition, as explained above, the government and the concessionaires have made particular agreements that have included changes in prices. Nowadays, the tolls are basically regulated through law.[6] On top of bilateral agreements, a 1990 national law established a general regulation for yearly price adjustments. This yearly adjustment is applied to all concessionaires in charge of national motorways. Initially, prices increased according to the following coefficient: $C = 0.95\Delta \mathrm{RPI}_{\mathrm{mean}}$, where C stands for change in price, and RPI stands for retail price index (in %).

However, since 2001 prices on national toll motorways[7] have been varying according to a price cap regulation. Tariffs are adjusted by the full increase of RPI minus a discount factor (X). The discount factor is constructed in such a way that its value rises with unexpected increases in traffic. Hence, unexpected increases in traffic reduce the extent of the tariff increase, within the bounds explained below. The regulatory system is formally constructed as follows:

$$T_t = C_R * T_{t-1} \tag{1}$$

where T stands for toll and C is such that

$$C_R = 1 + \Delta RPI_{mean} - X \tag{2}$$

X is defined as follows:

$$X = (1/100)\,[(ADT_{actual} - ADT_{predicted})/ADT_{predicted}] \tag{3}$$

where ADT stands for average daily traffic and $ADT_{predicted}$ refers to the ADT included in the economic and financial plan for the concession as approved by the Government Representation in the Concessionaire. In addition, X is bounded as follows:

a. As a general rule, X is bounded between 0 and 1 ($0 \leq X \leq 1$).

b. With regard to concessions that were already in effective operation before January 1, 1988, X is not bounded as in (a). Instead, the bounding rule works as follows:

$$1.15\Delta RPI_{mean} \geq \Delta RPI_{mean} - X \geq 0.75\Delta RPI_{mean} \tag{4}$$

In applying this regulation there is no consideration for features such as quality of service, maintenance, or the construction of new lanes. The price cap system is an attempt to link price changes with the actual evolution of traffic. As stated in Law 14/2000, the objective is to link extraordinary profits with reductions in the real prices of tolls, to share unexpected profits between users and concessionaires. In this way, it is worth noting that profits of the Spanish concessionaires increased substantially in the late 1990s due to the strong record of traffic in the toll motorways. An increasing discomfort with tolls in the territories where they are charged and the high profits of the concessionaires have motivated the enforcement of price ceilings.

However, older concessions are less constrained by the price cap regulation. There cannot be real increases (that is, above ΔRPI) in tolls in the concessions that began operating after January 1, 1988. In this way, X cannot take a negative value. Additionally, the maximum increase is $(1+\Delta RPI_{mean})$. For the older concessions, the maximum increase is $(1+1.15\Delta RPI_{mean})$, thus allowing real increases in price. With regard to the lower bound, comparison is not straightforward but still possible. Given that $X \leq 1\%$ for recent concessions, it is easy to see that $0.75\Delta RPI_{mean>(\Delta RPImean} - X)$ if $RPI_{mean} \leq 4\%$. Even if RPI can potentially go over 4%, it is not likely to happen. The European Central Bank sets the EU inflation target at 2%, and since the mid-1990s ΔRPI has been regularly below 4% in Spain.

Finally, let us note two paradoxes involved in this regulatory dynamic:

1. Part of the extraordinary increase in profits during the last years is derived from the conditions included in the re-negotiation agreements promoted by the government in the late 1990s. In fact, huge traffic increases are due to the conjunction of economic growth and reductions in tolls (given in return for concession extensions).

2. The response by the government has been to establish a price regulation that works as follows: the largest toll increases take place with the lowest traffic increases, whereas the lowest toll increases are associated with the largest traffic increases, in short, exactly the opposite of what efficient price regulation would advise – increasing prices with congestion.

IS IT POSSIBLE TO HOMOGENIZE THE MOTORWAY FINANCING MODEL IN SPAIN?

As explained above, even if some new concessions were franchised (occasionally with public subsidies) by the former government (Partido Popular), its policy maintained the mixed funding model. Till now, the new government (Socialist Party) that took office in mid-2004 has not made clear a policy model concerning tolls. Although no policy against the currently operating toll motorways is expected, there will likely be a downsizing of proposals of new toll motorways. Within this framework, the main topic of discussions among territorial governments, private sector, professionals, and scholars is whether there should be a homogenization of the motorways funding system.

The lack of homogeneity in the motorway network in Spain creates some deficiencies in the management and financing of the network and causes territorial inequalities that provoke increasing instability. The functional homogeneity of the major motorway network could allow implementing of more rational road policies, which would put an end to the high territorial diversity in financing models. Two basic alternatives could help to homogenize the network and remove the territorial inequalities and competition distortions that tolls impose on high-capacity roads: (1) generalizing tolls throughout the motorway network and (2) eliminating tolls.

Generalizing tolls: By the end of 2003 the total kilometers of motorways in Spain was around 12,000. Around 9,500 of them were operated without tolls. The practical feasibility of establishing tolls on free motorways has been appraised in several studies. Zaragoza (1992) points out that the material costs of establishing tolls would be very high. Also, he casts doubts on the legal feasibility of establishing tolls in many sections of free motorways that do not have a free road as an alternative. Soriano and Martín (1998) analyze the practical feasibility of this option taking into account the technological advances in charging tolls, and they infer similar conclusions. Indeed, establishing tolls can be expensive in terms of both time and money. Given the high costs and political difficulties involved in generalizing tolls, policy proposals in this direction are unlikely. The ideological stand of the party in government does not seem to matter in this regard.

Eliminating tolls where a free motorway alternative does not exist: This option has the advantage of being the most efficient as long as congestion does not exist. Hence, it applies especially to interurban sections where motorway capacity is high enough to absorb traffic coming from congested alternative roads. It has the disadvantage of requiring financial resources that could be invested in alternative projects. Still, it could be a sensible option from a financial point of view. This would require a gradual transition and the substitution of alternative tools that effectively make all users in Spain pay for motorway services. Furthermore, it would decrease the need for investment in currently congested roads parallel to toll motorways.

A question that is worth considering within this context is the effect of this kind of policy on the relationship between the public and private sectors, and particularly on the involvement of the private sector in the financing and management of infrastructure. As we have seen above, some Spanish motorway holdings companies (especially Abertis) have become global players in this business, and national policies are likely to regard them as a valuable asset for the overall Spanish economy. Let us analyze both issues separately:

1. *Private financing of infrastructure*: In this regard, it is important to note that direct tolls paid by users are not the only available form of private financing. In fact, there is also private financing when the public sector uses the model of postponed payment for an element of infrastructure. With respect to the "who pays?" question, there are systems other than tolls to make users contribute to infrastructure financing: periodic tariffs, specific tariffs on products such as combustibles that are closely linked to the use of the infrastructure, etc.
2. *Private management of infrastructure*: The private sector is in charge of operating and maintaining infrastructure, whether revenues come from the public budget or users (direct or indirectly). The schemes for cooperation between public and private sectors are diverse in this field.

This distinction is quite useful. Although usually forgotten, in the near future the financial requirements for maintaining and upgrading existent motorways will be higher than the financial requirements for investment in new motorways. Currently, maintenance of 80% of the Spanish motorway network relies on the public budget, and degradation has accelerated in the last years. A scheme of financing new investment through the public budget and maintenance through user charges could be convenient. This scheme is more demanding of users than current expectations. Recall that 84% of the new motorways (in the most recent national plan) are to be financed by the public budget, and users will finance only 16%. In contrast, the scheme proposed here is that all users contribute to financing the maintenance of all motorways.

This proposal would allow several options for cooperation between public and private sectors in the operation and maintenance of the national motorway network in Spain. One option is competitive tendering for time-limited management concessions, which could be financed either by a private operator charging tolls directly to users or through government payments from other resources.

Is it possible to incorporate this financing model in Spain if direct tolls must be used? Theoretically, yes. However, we have argued above that there is no likelihood of generalized direct tolls, as recent long-term government plans have shown. Other forms of user financing may be more useful and viable in moving toward a generalized and homogenized financing system in Spain. The lack of rationality in the Spanish motorway system has arrived at a point where effectively increasing homogeneity is more important than worrying about the sort of financing used. After all, economics usually deals with second best scenarios.

NOTES

1. It is remarkable that in a country like the United States, so much oriented toward private initiative, only two private roads were built during the 20th century (Engel, Fischer, & Galetovic, 2002). More recently, one other southern EU country, Portugal, has franchised private toll motorways. In addition to this, the major Italian franchisee of motorways, Autostrade, was privatized in 2000.
2. Bel (1999) contains a full account of the financial effects on the Treasury of the early motorways concessions.
3. Fernández, Molina, and Nebot (1983) suggest that the real business was in the building, as happened with the Spanish railways in the 19th century. The joint effect of tax and financial clauses in the concessions, along with commercial clauses – especially those allowing firms with stakes in a concessionaire to get involved in construction – is consistent with this hypothesis.
4. Matas and Raymond (1999) find negative and significant price-elasticities of demand in the Spanish motorways.
5. Sacyr-Vallehermoso has been traded in the Madrid stock exchange for many years. However, the traditional major activities of this holding have been construction and real estate development.
6. In this way, the government is not free to change the formula of tariff adjustments, because it has to follow the rules established by law. Therefore, since 1990

successive laws have established the regulatory framework for tolls. It has to be added that, within this framework, the government and a concessionaire can arrange bilateral agreements concerning tolls.

7. Tariff adjustments for regional motorways are ruled on through regional laws.

ACKNOWLEDGMENTS

The Spanish Commission of Science and Technology supports this research (CICYT, BEC2003-01679). Comments from Daniel Albalate, Andrea Greco, Peter Mackie, Giorgio Ragazzi, John-Hugh Rees, and Werner Rothengatter have been very useful. The usual disclaimer applies.

REFERENCES

Bel, G. (1999). Financiación de infraestructuras viarias. La economía política de los peajes. *Papeles de Economía Española, 82*, 123–139.

Engel, E., Fischer, R., & Galetovic, A. (2002). A new approach to private roads. *Regulation, Fall*, 18–22.

Fernández, R., Molina, E., & Nebot, F. (1983). El fracaso de la política de las autopistas de peaje. *Información Comercial Española, 594*, 37–54.

Gómez-Ibáñez, J. A., & Meyer, J. R. (1993). *Going private. The international experience with transport privatization*. Washington, DC: Brookings Institution.

Matas, A., & Raymond, J. L. (1999). Elasticidad de la demanda en las autopistas de peaje. *Papeles de Economía Española, 82*, 140–165.

Ministerio de Fomento (2003). *Anuario Estadístico 2002*. Madrid: Ministerio de Fomento.

Ministerio de Fomento (2004). *Anuario Estadístico 2003*. Madrid: Ministerio de Fomento.

SDC. (2003). *Informe del Servicio de Defensa de la Competencia N-03030 Sacyr-Vallehermoso/ ENA*. Madrid: Ministerio de Economía.

Soriano, F., & Martín, G. (1998). Instalar peajes en las vías rápidas. *Estudios de Construcción Transportes y Comunicaciones, 78*, 45–52.

World Bank. (1962). *Informe del Banco Mundial. El desarrollo económico de España*. Madrid: BIRF.

Zaragoza, J. A. (1992). La convergencia económica y la financiación de infraestructuras. *Información Comercial Española, 710*, 71–82.

PRICING AND FINANCING TRANSPORT INFRASTRUCTURES IN SWITZERLAND. A SUCCESS STORY?

Roman Rudel, Ornella Tarola and Rico Maggi

INTRODUCTION

With the publication of the White Paper 'European Transport Policy for 2010: Time to Decide' in 2001, the Swiss approach to regulating and financing the transport system was definitely introduced in the European transport policy debate as a kind of textbook example, the combination of the taxation of the heavy vehicles and the financing of new railway infrastructures conferring a particular appeal to the Swiss scheme. In spite of its relative simplicity, the present scheme results from a rather complex policy process, which has been shaped by the geographical position of Switzerland, the environmental concern of the Swiss population, and the actual functioning of the direct democracy inter alia.

During the 1980s and 1990s, Switzerland, at the centre of the Alpine region and not a member of the European Union (EU), came more and more under political pressure, due to the growing integration of the EU and the tremendous increase in road freight traffic on the north–south corridors across the Alps. The European neighbouring countries required from

Procurement and Financing of Motorways in Europe
Research in Transportation Economics, Volume 15, 205–213

ISSN: 0739-8859/doi:10.1016/S0739-8859(05)15016-1

Switzerland to lift its 28-tonne weight limit for heavy freight vehicles, to abandon gradually the restriction to circulate during the night hours and Sunday in order to reach the conditions to circulate in the EU. During the same period a grass-root movement started to highlight the environmental impact caused by the growing freight traffic on the road and focused the public attention to the unbearable impacts for the sensitive ecosystem of the Alps. The movement – the so-called Alpeninitiative – started a popular initiative with the scope to reduce the transalpine freight traffic to about 650,000 heavy vehicles per year – half the number of vehicles in 2002 – across the Swiss Alps by the year 2007 and to prohibit the construction of new highways in the Alpine zone.[1] While this measure was in clear contrast with the requirements of the European countries, the Swiss government, within the frame of a direct democracy, tried to find a compromise between national interests and European requests, promising to construct a new railway infrastructure across the Alps, to lift the 28-tonne limit, and to introduce the mileage related heavy vehicle fee. This compromise is supposed to link different policy goals, namely to shift freight traffic from road to rail, to reduce environmental impact and create new railway infrastructure (Rossera & Rudel, 1999).

In this chapter, we briefly present the Swiss financing and regulation scheme of transport system (see next section) as it worked until 2001. Then we focus on the relationship between Switzerland and the EU and the emergence of the transalpine freight problem, which inspires the realized changes in the traditional pricing scheme. The subsequent section deals with the present pricing regime and the new policy target, according to new rules, which have been introduced in January 2001. Finally, some conclusions on the impact of the new pricing regime and its potential to shift freight traffic from road to rail are drawn.

THE SWISS FINANCING AND REGULATION SCHEME UNTIL 2001

The origin of the highway network is based on a popular vote in 1958 in favour of the construction of the highway system and the way to finance it. The federal government has the task to create the highway system, while cantons are in charge of the construction and maintenance of the highways according to national obligations. The network has reached at the end 2002 more than 1,300 km (excluding 305 km of two-lane highways and 96.5 km mixed traffic national roads). Yet it has still do be considered uncompleted, as the lacking pieces constitute about 10% of the overall system.

The highways in Switzerland are entirely public. The construction, the maintenance as well as the improvements of the highways are in charge of the regional governments. The main financial revenues are the fuel tax and the vehicle tax. The fuel tax and heavy vehicle tax revenues flow in the general budget of the federal government. Seventy percent of these tax revenues are earmarked for the construction, operation, and maintenance costs of the road system at the national and cantonal level and only a very small fraction is devoted to the communal roads (Blöchlinger, 1999). Up to 90% of the infrastructure costs at the communal level are in charge of the local government. Further financial means are levied on the light vehicles; these taxes on passenger cars are collected and used at the cantonal level. These revenues even finance projects having an indirect relation to the road network.[2]

The introduction of a heavy weight vehicle fee in 1978 for national trucks over 3.5 tonnes, represents the first financial regulatory instrument in the Swiss freight transportation market. The tax was differentiated on the basis of the vehicle weight ranging initially from 800 CHF to 1300 CHF. The foreign haulers paid a fixed charge. The cost for crossing Switzerland from Basle to Chiasso (300 km) for foreign trucks was about 40 CHF. This amount had to be considered as extremely cheap, compared to the passages at the Brenner or Mont Blanc. However, this price was applied under the 28-tonne limit scheme. A major consequence of this regime was a considerable traffic flow of empty trucks, deviated from the Brenner and Mont Blanc route. Finally, in 1994 the so-called 'Autobahnvignette' (highway tag/batch) was introduced. It is a low flat tax consisting of an annual permit for driving on highways[3] and representing the main regulatory instrument for private cars.[4]

Until the mid-1990s the expenditure for the highway network went in parallel with the railway infrastructures, rapidly increasing since 1998, when the construction of the new railway infrastructure across the Alps (Lötschberg and Gotthard) started[5] (Fig. 1). In spite of this increasing pattern of the railway infrastructures expenses, the railway infrastructures investment remains lower than the road infrastructure investment.

The investments in the 1990s in the railway infrastructures represent the political will to balance the investment flows again in favour of the railway. The main purpose was to update an infrastructure generally older than 80–100 years. While the Swiss population voted in 1987 in favour of the project Rail 2000, the global investments in road infrastructures remained during the last decade always higher than for railway (Carron, 2003). Further, since 1997 the balance between expenses and revenues has

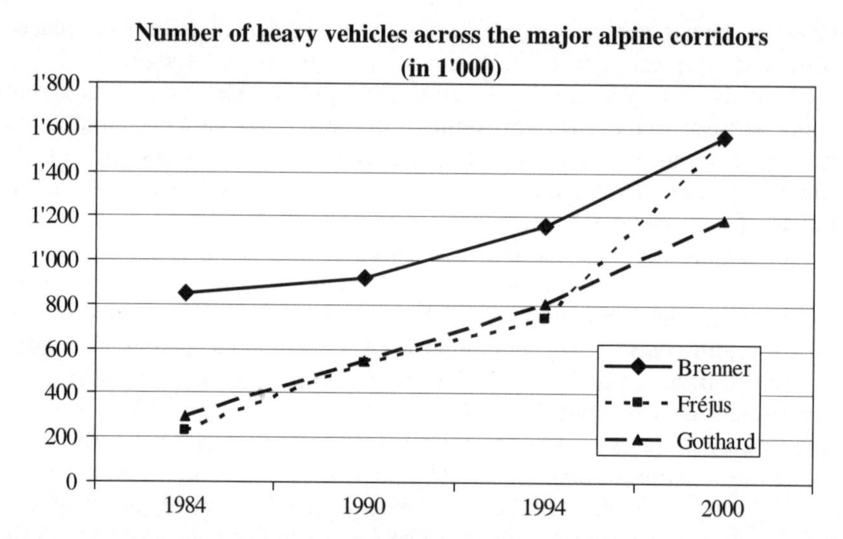

Fig. 1. Traffic on Alpine Corridors. *Source:* Alpinfo, Swiss Federal Office for Spatial Development, Berne, 2004.

had a positive sign, indicating that the road system, contrary to the railway, is financed on the revenues generated by the road traffic itself. In spite of this positive trend, a strong debate is taking place between the EU and Switzerland on the economic and environmental consequences of transalpine freight traffic, induced to strongly reconsider the traditional Swiss approach to the transport system. In the following paragraphs, we first focus on the transalpine freight problem as it emerged in the 1990s and then we describe the main changes, which have been introduced in the freight traffic regulation.

EU–SWITZERLAND AND THE EMERGENCE OF THE TRANSALPINE FREIGHT PROBLEM

The geographical situation and the dimension of Switzerland are relevant for the Swiss regulation approach and the freight transport market. The national railway company yielded considerable rents due to their position on the shortest north–south connection across the Alps. At least until the mid-1970s, the Gotthard corridor represented a real cash cow for the railway company. The road regulation with the 28-tonne limit and the prohibition to circulate during night hours contributed essentially to this position. Certainly, it helped to maintain the high market share of rail and intermodal freight transport to a considerable extent until now. Indeed, one of the most prominent characteristics in the transalpine freight market is the high share of rail transport in Switzerland compared to France and Austria. While in France and Austria more than two-thirds of freight transport is represented by road transport, in Switzerland the share is reversed. The favourable position of the railway was taken for granted and many policy makers were blind for the rapid changes in the transalpine freight market. As long as the freight traffic flows were expressed in term of volume, it was difficult to recognize the rapid changes in the freight market. Yet, the opening of the highway tunnels across the Swiss Alps in 1980, intended to promote the private passenger mobility, became a major driving force of these changes (Ratti and Rudel, 1993). In 1980, only a few thousand lorries passed the new highway tunnel. This figure rapidly increased up to over 1.2 million lorries a year in 2000.[6]

In spite of this evolution, the EU put pressure on Switzerland to open up the 'transit corridor' and to loosen the severe restrictions on heavy road traffic (Maggi, 1991), also with the aim of reducing the environmental burden of freight road traffic thanks to an enhanced productivity. The majority of the Swiss population, however, conceived open road freight traffic along the shortest transalpine route as both an invasion and an ecological disaster.[7] In 1992, the Swiss population voted in favour of the Alpeninitiative, to reduce road freight traffic and to build two new railway base tunnels across the Alps.

With the vote of the Alpeninitiative, the Swiss government realized that it would be politically untenable to loosen the restrictive regulation in the transalpine market. However, the government was forced to find a way to reduce the pressing European request and to offer an alternative to the discriminatory limitation proposed by the Alpeninitiative. In the negotiations to the bilateral agreements between Switzerland and the EU, the countries found a mutually satisfactory solution. The cornerstones of the agreement are the construction of the new railway infrastructure (NEAT), the gradual abandoning of the 28-tonne limit and the introduction of a mileage related heavy vehicle fee (LSVA).

THE PRESENT PRICING REGIME AND THE TARGET OF SWISS TRANSPORT POLICY

In June 2002, the land transport agreement, part of the overall agreement between Switzerland and the EU, came into force. The transport policy is essentially an answer to the rapidly growing freight traffic on the road. In Switzerland, the problem of freight traffic is a prominently international one and a key factor in the relationship with the European neighbour states. The solution adopted in the agreement mainly signs a shift from a regulation Scheme (28-tonne limit, low fixed transit charge) to a scheme compatible with the European transport policy. It particular, the EU has recognized the goal of the Swiss transport policy. Moreover, both parties agreed on the introduction of a heavy vehicle fee (for vehicle with a total weight over 3.5 tonnes passenger and freight) on all Swiss roads (Suter, 2002). The introduction is neither limited to the highway system nor to the transit corridors.[8] Domestic and foreign vehicles are treated in the same way and the fees comply the non-discrimination principle.

The new regime is introduced and implemented according to the steps given in Table 1.

The setting of the fee is somewhat ambiguous. The fee is supposed to be fixed on the rationale of external cost pricing based on studies in 1998. The initial fee was about twice the present fee after the negotiation process between Switzerland and EU. In the land transport agreement it is clearly stated that the fee is set in a way that the resulting fee for a transit of a heavy vehicle form Basle to Chiasso (about 300 km) should not exceed the price of CHF 325 or approximately 200 euros (ARE, 2002). This transit price is comparable to the fees currently applied on the Fréjus (217 euros for 346 km) and slightly higher than those on the Brenner corridor (105 euros for 335 km).

We show the difference between the flat rate fee of the 'old' regulation scheme and the new mileage related fee introduced in January 2001 in Table 2.

Contrary to the sophisticated Mautsystem in Germany based on satellite technology the heavy vehicle fee in Switzerland requires only a simple

Table 1. Fees and Weight Limits.

	Before 2001	2002–2004	After 2004
Land transport agreement		In force	In force until 2013
Weight limit (in tones)	28	34	40
Heavy vehicle fee (CHF/tkm)	Fixed fee	0.017	0.025
NEAT			2006/07 opening of the Lötschberg 2012/13 opening of the Gotthard

Source: Adaption from ARE (2002).

Table 2. Flat Fee compared to Mileage Fees.

	Flat Rate Fee in the Year 2000	Heavy Vehicle Fee
Rate of fee	CHF 1,300–1,800/year; Euro 870–5,300/year (depending on weight class)	Maximum 2.75 cts/tkm; 1.8 cents/tkm
Staggering	According to weight class	According to: weight class; emission category (Euro norm); distance traveled
Transit price; Basle–Chiasso	CHF 40/day; Euro 25 (daily fixed rate for all weight classes)	Maximum CHF 325; 200 euros

Source: Swiss Federal Office for Spatial Development, Berne, 2002.

on-board unit connected to the tachograph in order to register the distance driven. Foreign truck drivers use a smart card recording the distance driven on Swiss roads at the border stations. The smart cards or the data from the on-board unit are sent once a month to the custom authority for the billing process. The collection of the electronic fee is therefore very simple and hardly interrupts the transport flows. Enforcement is through customs checks at the borders, roadside checks and checks in the accounts of Swiss haulage companies (Perkins, 2004).

The revenues of the heavy vehicle fee accounted in 2002 for about CHF 770 million (500 million euros) and are likely to remain of this magnitude for the next 20 years. The fee is paid by foreign as well as be Swiss haulers: two-thirds of the revenue directly flows into a special investment fund for financing the new transalpine railway infrastructure (FinÖV), as well as three other major railway projects. Further financial resources of the special fund stem from earmarked revenues of the fuel tax, 0.1% of the VAT and the rest is covered with a loan granted by the Swiss Confederation.

CONCLUSIONS

The Swiss approach to pricing freight traffic and financing new infrastructures has gained considerable popularity among policy makers on the national as well as the European level. The heavy vehicle fee was designed to reach multiple goals: to generate a handsome revenue earmarked for the construction of the new railway infrastructures across the Alps, to favour the modal shift from road to rail and to reduce the environmental impact in the Alpine regions (Flyvbjerg, Bruzelius, & Rothengatter, 2003).

A closer look at the 'genesis' of the different interplaying elements clearly demonstrates that the present policy regime is the outcome of a long and complex process with different stakeholders and parties involved. The first effect of the new pricing scheme with a weight- and emission-dependent fee has clearly been to induce the road haulage sector to renew its fleet and adapt it to the new conditions. A similar change in the composition of the fleet of lorries could be observed with the introduction of the eco-point system in Austria. A further effect is concerned with the internal transport industry and its reaction to the sharp increase in the road charges. In general, the major transport costs could be compensated by the higher productivity of trucks due to the lift in the weight limit. In 2001 and 2002, the long-term trend in the constantly growing number of road freight traffic has been broken as a consequence of the new charging scheme. The effect on the modal shift is less evident. However, the figures and available data seem to support the Swiss transport policy so far.

Yet, the preliminary effects of the new pricing scheme are not sufficient to eliminate the serious doubts on the efficiency of the new freight traffic policy. Increasing congestion problems on the Swiss highway system, other than on the north–south corridor across the Alps, are rapidly increasing and urban transport requires new investments. Moreover, the question of the connection of new railway infrastructure with Italy still remains to be answered and the modal shift is limited by terminal infrastructure capacities as well as low service quality of railway transport (Truffer et al., 1998).

NOTES

1. This applies in particular to the Gotthard highway tunnel, a serious bottleneck in the north–south corridor, especially during holidays and weekends. The request to double the two-lane tunnel for safety reasons has been rejected recently in a popular vote.

2. Some are designed to protect the environment and the landscape while others to support the combined freight traffic.

3. Contrary to most European countries, using the Swiss highway network is based on an annual tax independent of the kilometres driven, the vehicle category, or energy consumption.

4. In spite of the relatively consistent levies in the road freight traffic and the environmental concern, various attempts to stronger regulation or limit private car use had no chance to be approved in a popular vote as the majority of the car drivers were the Swiss population.

5. The Gotthard railway tunnel constitutes the longest tunnel worldwide. See also www.Alptransit.ch
6. The rapid growth of road freight traffic across the Swiss Alps also was in line with the major crossings in France and Austria.
7. Forecasts (Graf, 1995; Vittadini, 1992) predicting a doubling of the freight traffic in less than 20 years supported this perception.
8. This is one of the main differences with the German 'Mautsystem'.

REFERENCES

ARE. (2002). *Fair and efficient. The distance-related heavy vehicle fee in Switzerland.* Berne: Swiss Federal Office for Spatial Development.

Alpinfo. (2004). Swiss Federal Office for Spatial Development, Berne.

Blöchlinger, H. (1999). Finanzierung des Verkehrs von morgen: Analysen und Reformen. *National Research Programme*, NPR-41, Berne.

Carron, N. (2003). *Politiques de financement et d'investissement en infrastructures de transport. Example de la Suisse.* Berne: Swiss Federal Office for Spatial Development.

Flyvbjerg, B., Bruzelius, N., & Rothengatter, W. (2003). *Megaprojects and risk. An anatomy of ambition.* Cambridge, UK: Cambridge University Press.

Graf, H.G. (1995). *Perspektiven des schweizerischen Güterverkehrs 1992–2015. Dienst für Gesamtverkehrsfragen.* Federal Department of Environment, Transport, Energy and Communications.

Maggi, R. (1991). Transport regulation in Switzerland. In: K. Button & D. Pitfield (Eds), *Transport deregulation. An international movement.* London: Macmillan.

Perkins, S. (2004). Charging for use of roads: Policies and recent initiatives. ECMT. *Fifth annual global conference on Environmental Taxation Issues, Experience and Potential*, Pavia.

Ratti, R., & Rudel, R. (1993). Tableau de l'évolution des transports dans l'arc alpin. *Revue de géographie alpine*, *4*, 11–26.

Rossera, F., & Rudel, R. (1999). The supply of combined transport services. Increasing their market penetration. *National Research Programme*, NPR-41, Berne.

Suter, S. (2002). Theoretical view on pricing. Latest developments in research: Theory, application and impacts. *Thematic network: Alp-net*, Berne.

Truffer, B., Rudel, R., Cebon, P., Dürrenberger, G., Jeager, C., & Rothen, S. (1998). Innovative social responses in the face of global climate change. In: P. Cebon (Ed.), *Views from the Alps. Regional perspectives on climate change.* Cambridge, MA: MIT Press.

Vittadini, M. R. (1992). *Il quadro della domanda di trasporto. Il trasporto di merci e di persone attraverso le alpi. Situazione e prospettive di evoluzione.* Milan: IRER.

FINANCING ROADS IN GREAT BRITAIN

Peter Mackie and Nigel Smith

INTRODUCTION

The British tradition of road finance and procurement has been one of almost complete separation between decisions on road user taxation and expenditure on roads. Road users pay taxes which are set by the Treasury alongside income and indirect taxes as part of fiscal policy. Expenditures on roads are undertaken by a mixture of the Highways Agency for national roads and local authorities for local roads.

It is worth noting two significant moments in history. In the early part of the twentieth century, road user taxation was earmarked for expenditure on roads, according to the so-called 'benefit principle' of public finance. In an economic crisis in 1926, Winston Churchill, the Chancellor of the Exchequer, raided the Road Fund and destroyed the link between road tax revenue and road expenditure forever. Then, during the 1950s, when the motorway network was being planned, the original expectation was that some motorways would be tolled. Subsequently, this policy was reversed, and the motorways were designed in an integrated way with the general road system and were toll-free (Charlesworth, 1984).

The principal sources of taxation from road vehicles are fuel duty and vehicle excise duty. Governments have adopted a rather loose policy that all classes of road users should pay taxes which at least cover their road use

Procurement and Financing of Motorways in Europe
Research in Transportation Economics, Volume 15, 215–229

ISSN: 0739-8859/doi:10.1016/S0739-8859(05)15017-3

costs. This was seen as an important principle for heavy goods vehicles, in order to assure 'fair competition' between road and rail-based freight transport. This led to engineering and economic studies of the cost structure of road provision and relationships with taxes (Ministry of Transport, 1968). For many years, until the mid-1990s, an annual report of road use costs and taxes was produced, though there remained many questions about vehicle categories, allocated cost formulae, average versus marginal costs, treatment of external costs and so on.

A relatively recent study gave the headline results as shown in Table 1 for 1998. The main points which can be inferred from Table 1 are as follows. If only road capital and operating costs are considered (rows 1 and 2) then all vehicle classes comfortably cover their allocated track costs. If, however, environmental and accident costs are added in, while cars continue to cover their fully allocated costs, for goods vehicles the outcome depends on whether low or high cost estimates are accepted. Moreover, as Sansom, Nash, Mackie, Shires, and Watkiss (2002) show, the comparison between marginal use cost including congestion and marginal revenue looks very different.

Table 1. Fully Allocated Cost and Revenue by Vehicle Class (Pence/Vehicle km) (1 Pence = 1.4 Euro Cents).

	Car		Light Vans		Heavy Goods Rigid		Heavy Goods Artic.	
	Low	High	Low	High	Low	High	Low	High
Cost of capital	0.70	1.21	0.83	1.43	1.45	2.49	1.88	3.22
Infrastructure operating cost	0.33	0.43	0.38	0.49	4.61	6.00	8.74	11.36
Accident cost	0.07	0.82	0.04	1.46	0.04	0.61	0.03	0.50
Air pollution	0.18	0.88	0.71	3.35	1.65	8.26	1.41	7.63
Noise	0.16	0.52	0.30	1.00	0.87	2.89	1.31	4.35
Climate change	0.12	0.47	0.18	0.72	0.44	1.74	0.71	2.86
Total cost	1.6	4.3	2.4	7.5	9.1	22.0	14.1	29.9
Vehicle excise duty	1.03	1.03	1.03	1.03	2.25	2.25	2.50	2.51
Fuel duty	3.86	3.86	3.86	3.86	13.1	13.1	14.4	14.4
VAT on fuel duty	0.68	0.68	0.68	0.68	2.29	2.29	2.53	2.53
Total revenue	5.6	5.6	5.6	5.6	17.6	17.6	19.5	19.5
R:C ratio	3.5	1.3	2.3	0.7	1.9	0.8	1.4	0.6

Source: Sansom et al. (2002), Tables 7.2 and 7.3.

IMPLICATIONS FOR PROCUREMENT OF ROADS

Given this background, until the 1990s, the conventional model of public funding of public roads on the public sector balance sheet was with few exceptions the only way of procuring roads in Britain. This is explained by a number of factors including:

- A very strong public sector procurement tradition; that is, public procurement from private construction firms using competitive tendering.
- Very little tolling of roads; therefore, few opportunities for privately owned toll-financed schemes.
- A very strict Treasury attitude to private finance (the Ryrie rules). These rules, in simple terms, were:
  - *No additionality*. Private finance for sector investment should replace public finance not be additional to it.
  - *Risk transfer*. Private finance is acceptable provided that genuine risk transfer to the private sector takes place.

Taken together, these conditions mean that there were few circumstances in which the system would be both incentivised and capable of taking advantage of private finance. For an extensive discussion, see Heald (1997), who quotes Chief Secretary of the Treasury John Major (1989) as saying: "The Ryrie Rules are thought to be incomprehensible, and to hamper private finance by setting impossible hurdles."

So, until about 10 years ago, private finance of road schemes was very unusual. However, there were exceptions.

First, there are important tolled estuary crossings. On these crossings, the alternative routes are generally poor and/or involve lengthy diversions, so that the price elasticities are relatively low. The allocative efficiency losses from tolls are small relative to those which would occur on the main inter-urban network. Two examples of tolled crossings are the Severn Crossing between Bristol and South Wales and the Dartford Crossing on the M25 to the east of London. In both cases, during the 1980s, these crossings were running out of capacity and investment in new bridges was required. The model chosen in both cases was:

- a franchise arrangement with a private consortium taking over the existing facility and providing a new crossing for a franchise period;
- reversion of assets to the government at the end of the period;
- regulated tolls;

- some risk transfer to the private sector, but a flexible franchise period permitting risk-sharing.

Broadly speaking, for these schemes, the ex ante planning risks were taken by government. The schemes were approved before the franchises were let. The construction and maintenance risks were taken by the consortia, with the traffic risk being shared between the government and the consortia by the device of the variable operating period. Both schemes have worked in the sense that the projects have been delivered and the consortia have survived. However, it must be stressed that these are very heavily trafficked toll crossings where commercial viability is not really a problem. The more interesting question is that of the public sector comparator and whether there was a net advantage in going private. It is believed that in the case of the Second Severn Crossing, the capital cost of the scheme was very similar to the expected public sector capital cost. However, an important issue is whether conventional public sector procurement has a greater tendency to cost overruns than private procurement: in other words, would the public sector actually have delivered the project at the ex ante expected public sector capital cost? Unfortunately, the counterfactual is not directly observable.

At the other end of the spectrum, private finance has also played a part in the procurement of road improvements via Section 52 of the Highways Act 1980 for central government schemes and Section 106 of the Town and Country Planning Act 1990 for local authority schemes. These arrangements come into play in the context of developer contributions as part of a larger deal involving office, retail or other developments. For example, the developer might contribute to improvements to the junction between the local access road and the main road to which the development is connected. From the perspective of the local authority, this may enable an improvement which was needed anyway to be brought forward. Clearly with this type of co-funding there are issues about impacts on local authority prioritisation. Control of procurement remains in the hands of the public authority; this is a private contribution to what remains a public scheme. There are, however, many difficulties in the UK with the concept of capturing planning gain where multi-developer sites are concerned.

So, the position at the end of the 1980s was:

- rare large schemes franchised to private sector consortia;
- developer contributions to local road improvements within the conventional public procurement process.

ROADS POLICY UNDER THE MAJOR GOVERNMENT (1992–1997)

In the early 1990s, a combination of events led to a serious squeeze on the roads sector. Economic recession, the collapse of the strong pound policy leading to withdrawal from the exchange rate mechanism and extreme reluctance to increase taxes were general influences. Within the roads sector, constraints included slow planning procedures, increased opposition from environmental interests and a loss of confidence in the economic case for some schemes. The treatment of induced traffic in appraisal came in for particular attention (Standing Advisory Committee on Trunk Road Assessment, 1994). All of this meant that the motorway and trunk road public expenditure budget fell by a third in cash terms (more in real terms) between 1993/94 and 1998/99, while the local roads budget halved.

In this environment, with a heavy squeeze on public funds, there was considerable interest in funding some projects privately. The private finance initiative (PFI) was a multi-sector initiative designed to support this. Basically, the government dropped the "no additionality" rule of Ryrie's. Private finance was acceptable provided there was genuine risk transfer to the private sector and the private sector deal was better than procurement via the public sector – the so-called comparator. At sector level, given the shortage of public capital, the system was incentivised to use private finance as a supplement to conventional public sector investment.

In the roads sector, with very few toll roads, and an overriding policy of ultimate reversion to public ownership, the chosen medium for introducing PFI into the roads sector was the shadow toll approach. Thus, private consortia were invited to enter a tender competition to design, build, finance and operate (DBFO) new roads, or in some cases reconstructed roads, and ultimately to transfer them back to government. Effectively, the winning bidder was the one who offered to take the franchise for the lowest shadow toll, in terms of pence per unit of traffic. In practice, there was a schedule of bids for each tranche of traffic and a minimum/maximum operating period. The winning bid had to have a lower NPV of cost to the public sector than the public sector comparator.

Why might going private result in a lower present value of cost? We are assuming here that like is being compared with like. In practice, this might not be the case – a private finance deal might bundle together a road together with planning consent on adjacent land. It will then not be surprising if the private sector bid for the bundle is lower than the public sector cost of building the road alone. This is a biased comparison. Leaving that to one side, the arguments lie mostly in the area of efficient risk transfer. This is the proposition that risk is efficiently located with the party who is best placed to manage the risk. If this is achieved, then all else equal, the present value of cost of a given scheme will be minimised.

Some examples in the roads sector are:

- *Design risk* – conventional public sector procurement relies on design and construction methods which conform with relevant design manuals. If the design risk is passed to the private consortium, then innovative approaches may be used. For example, in the M1/A1 link road scheme to the south of Leeds, innovative methods were used to dig the tunnels at the major intersection with the M62 resulting in cost efficiencies and many months fewer on site.
- *Construction risk* – a conventional contract relies on an adversarial procurement process with claims engineers acting on behalf of the contractor and other claims engineers on behalf of the client. If the construction risk is transferred, this adversarial process is internalised within the construction company with significant efficiency gains.
- *Operating risk* – a conventional contract gives the contractor no incentive to choose the optimum quality. A DBFO scheme stimulates a whole-life costing approach because the capital and future maintenance costs are the responsibility of the same agent.

An overarching argument is that in a DBFO-type world, scrutiny is more likely to be effective by all parties. The market forces the issues within the project to be confronted more effectively than a bureaucratic decision process. Under DBFO, post-contract specification changes will be expensive, so the risk of politically induced specification changes is reduced. The bidding process itself encourages efficiencies provided conditions for a fair auction exist.

As against this, there are some factors which might make the present value of costs higher under private procurement than through the conventional route. These are:

- *Inappropriate risk transfer* – if for some reason, the wrong risks are transferred to the private consortium, cost will be increased. For example, in the case of the Birmingham Northern Relief Road (M6 Toll) much of the planning risk was transferred to the consortium. These were risks which the consortium was ill-placed to manage, which caused huge delays, and which nearly led to the project becoming unfinanceable. Again, there are arguments that transferring traffic risk is inappropriate. If traffic volumes depend primarily on economic growth, fuel prices and other macro variables, most of the risk is not amenable to management by an individual private consortium. Inevitably, they will require a risk premium to compensate them for taking on an unmanageable risk.

- *Inappropriate attitude to risk* – if the agents are extremely risk averse, then it may be impossible to form a market, or certainly to beat the public sector comparator.
- *Differential cost of capital* – this is controversial, but if real net of tax private sector costs of capital for procurement of a given asset in a given way are inherently higher than public sector costs of capital (to cover bankruptcy risk etc.), then the efficiencies outlined above need to be sufficiently large to overcome the cost of capital premium. This leads to many arguments about whether the public sector discount rate adequately represents the cost of capital of procuring a particular asset using public funds (Currie, 2000; Heald, 2003).

THE HISTORY OF DBFO PAYMENT MECHANISMS[1]

The first PFI roads were rolled out in 1994. By 2002, 14 DBFO roads had been completed, and these are summarised in Table 2. They are concession contracts governing construction works and operation and maintenance commitments for 30-year (maximum) terms. The first generation payment mechanism incorporated traffic usage (the shadow toll component), service availability and scheme performance as follows:

- Bidders specified 'bands' of traffic that would attract lower payments as traffic volumes increased. The top band generated no additional return for the concessionaire, so that the procuring agency's financial exposure was capped.
- Traffic was divided into vehicles below 5.2 m in length and above 5.2 m in length, as a proxy for light and heavy axles to reflect differential maintenance costs.
- The service availability component was designed to incentivised contractors to complete construction works on time or early.
- Scheme performance reflected lane-change charges and highway safety considerations.

Table 2. DBFO Road Schemes to Date.

Scheme	Contract Award	Concessionaire (DBFO Co.)	Cost (Mil £)	Summary Description
Highways Agency Schemes				
Tranche 1				
A69 Newcastle – Carlisle	January 1996	Road Link Ltd.	9	Project length = 84 km; construction of a 3.5 km bypass
A1(M) Alconbury – Peterborough	February 1996	Road Management Services Ltd.	128	Project length = 21 km; motorway widening
A417/A419 Swindon – Gloucester	February 1996	Road Management Services Ltd.	49	Project length = 52 km; construction of three new sections of road
M1–A1 Motorway Link	March 1996	Yorkshire Link Ltd.	214	Project length = 30 km; construction of new motorway, motorway widening and new interchange
Tranche 1A				
A50/A564 Stoke – Derby Link	May 1996	Connect Ltd.	21	Project length = 57 km; construction of a 5.2 km bypass
A30/A35 Exeter – Bere Regis	July 1996	Connect Ltd.	75	Project length = 102 km; Construction of two new sections of road and a 9 km bypass
M40 Denham – Warwick	October 1996	UK Highways Ltd.	65	Project length = 122 km; motorway widening
A168/A19 Dishforth – Tyne Tunnel	October 1996	Autolink Concessionaires Ltd.	29	Project length = 118 km; on-line widening
Tranche 2				
A13 Thames Gateway[a]	April 2000	Road Management Services Ltd.	146	Project length = 24 km; on-line upgrade and improvement schemes

Table 2. *(Continued)*

Scheme	Contract Award	Concessionaire (DBFO Co.)	Cost (Mil £)	Summary Description
A1 Darrington – Dishforth	September 2002	Road Management Services Ltd.	240	Project length = 22 km; construction of two new sections of motorway and communications
Scottish Office Schemes				
M6/A74	December 1996	Autolink Concessionaires Ltd.	96	Project length = 90 km. Construction of new sections of motorway and trunk road (for non-motorway traffic)
Welsh Office Schemes				
A55 Llandegai-Holyhead	December 1998	UK Highways Ltd.	120	Project length = 50 km. Construction of section of trunk road
Local Authority Schemes				
A130 (A12–A127)	October 1999	County route	75	Project length = 15 km. Chelmsford bypass
Newport Southern Distributor Road	June 2002	Morgan Vinci	50	Project length = 9.3 km. new crossing of the R. Usk
Total Capital Value			£1.3 billion	

[a]In July 2000, project responsibility passed from the Highways Agency to Transport for London (TfL).

Under this scheme, construction risk, operating risk and a significant proportion of traffic risk was passed to the concessionaire.

Subsequently, the system has moved in two respects. The traffic usage related component has been first reduced in significance and then eliminated. In other words, it has been concluded that passing traffic risk to the concessionaire is not compatible with the principles of risk assignment described above. Second, the service availability concept has been strengthened to include the condition of the road and defined to differentiate between peak and off-peak periods.

The most recent payment mechanism (the active management payment mechanism) has arguably more in common with the facilities management emphasis found in other PFI sectors than with earlier DBFO roads. Payments reflect service management, safety performance and a new component, congestion management, which is now the principal driver of the payment mechanism. The congestion management formula is complex and seeks to distinguish between factors which are and are not within the concessionaire's control. Bain and Wilkins (2002) suggest that this introduces a new set of risks from the concessionaire's point of view relating to performance forecasting accuracy, scrutiny and monitoring accuracy, and the ability to anticipate contingent events over a 30-year contract period. In short, there are anxieties that this is too sophisticated and will need to be paid for by an enhanced risk premium.

NATIONAL AUDIT OFFICE REVIEW OF THE FIRST FOUR DBFOs

This report (National Audit Office, 1998) is probably the most detailed assessment of the DBFO approach to road procurement in the UK. Some key findings are given below:

- In two of the four cases, DBFO was clearly preferable to the public sector comparator (see Table 3). These were the more capital intensive projects.
- The choice of discount rate for the public sector comparator is critical to the calculations. The NAO considered that the 8% rate was too high.
- The Department for Transport should consider the relative priority of road projects when selecting projects for future rounds of DBFO tenders. This is clearly important in principle if there is insufficient capital to carry out all desirable projects.
- There is a need to stimulate competition in advance and, for example, by permitting variation in technical requirements.
- There is a need for effective measures to secure contract compliance.
- Correct payment requires accurate and reliable traffic measurement, and audit of complex financial calculations.
- Compared to traditional procurement, the process was time consuming

Table 3. NPV (£ million) of Public Sector and DBFO.

Project	Public Sector (NPV)	DBFO (NPV)	Difference
M1–A1	372	288	84
A1(M)	222	192	30
A419/A417	137	140	(3)
A69	66	78	(12)
M40	329	228	101
A19	211	171	40
A50/A564	91	83	8
A30/A35	161	180	(19)
Total	1,589	1,360	229 (−15%)

Table 4. Net Value of PFI Compared to Public Sector.

	M1–A1	A1(M)	A419/417	A69
Net value of private finance relative to public sector comparator				
At 8%	112	50	11	−5
At 6%	84	30	−3	−12

> and bidding costs were high. The cost to all parties of letting the first four contracts was high compared with conventional road schemes. As with many tendering regimes, the higher transactions costs need to be offset against the efficiency gains.

Subsequently to the NAO Report, the House of Commons Select Committee on Transport reviewed the performance of the first eight DBFO schemes (HC 844, 1998). These schemes – a mixture of new build and major reconstruction projects – were on average 15% cheaper than the public sector comparator assuming a 6% public discount rate and using standard national traffic growth assumptions (see Table 4). It is noteworthy that the schemes are in two groups – the four more capital intensive schemes offering 20–30% savings and the others offering small or negative savings.

SINCE 2000

Since about 2001, the use of private finance for public roads has fallen out of favour, and we have reverted to a binary model with the use of private finance for (a very few) toll roads and public finance for free roads. Certainly, tranche 3 of the DBFO roads programme was cancelled. Why is this?

First, and most important, was the realisation that DBFO is really only a way of converting an up-front capital sum into a mortgage over the life of the scheme. Unless the road is tolled, it essentially re-profiles the public sector costs. By 2001/02, mortgage payments of around £200 million/year on existing DBFO schemes were going to be required for the foreseeable future. This was a significant proportion of the Department of Transport's available budget for new schemes and is referred to within the department as 'silting up'. This is akin to saying that, without private funding through tolls, there is ultimately no escape from the first Ryrie rule, 'no additionality'. In this case, it is preferable with some exceptions, to retain the whole roads budget within the public sector. This at least avoids the danger of giving excessive priority to schemes which happen to be 'private finance-feasible' over other schemes.

Second, there is increasing evidence that, in general, private finance is the more expensive method of procuring roads. In a very extensive study, though hampered to some extent by commercial confidentiality, Edwards, Shaoul, Stafford, and Arblaster (2004) concluded that "The SPVs' total effective cost of capital was about 11% in 2002. Although the NAO believed that this additional cost of private finance (6 percentage points above Treasury stock) represented the cost of risk transfer, it was difficult to see what risks the companies actually bore since their payments were guaranteed by the Government and based on shadow tolls, which in the context of rising traffic meant that they were insulated from downside risk at the Highways Agency's expense."

Third, and more dubiously, the reduction in the public sector discount rate from 6 to 3.5% in 2003 makes private finance even less attractive relative to the public sector comparator. However, the low discount rate coupled with an extreme shortage of public capital for roads poses an appraisal conundrum. While 3.5% real may be a sensible rate to use to reflect social time preference (i.e. the relative present and future value of £1 worth of resources), it is certainly not a reasonable value for the social opportunity cost of £1 worth of free resources in the hands of the Treasury. Given the acute scarcity of public capital, there is a clear need to use a shadow price of public funds, both present and future, in prioritising public investment, but for some reason the UK Treasury has been extremely reluctant to acknowledge this. As a result, even projects with potential efficiency gains are unlikely to be able to overcome the public/private cost of capital differential.

Where does that leave the UK? In terms of the procurement of roads, even if DBFO roads are currently dead, there are lessons to be learned about conventional procurement practice. There is a market to be exploited in innovatory construction practices. There is probably a market advantage in private sector control of projects, including elimination of wasteful adversarial practices regarding claims and contingencies. There are probably advantages to be had by internalising the whole life costing culture within construction incentives. It should be possible to gain these benefits through some form of design, build, indemnity and transfer contract without incurring the costs of private finance. In other words, the benefits of bringing the private sector into roads may come more in the form of management than in the form of finance. The policy question is how to achieve these benefits without a full DBFO form of contract.

In terms of road user charging, there is more policy activity in Britain than for many years. The Government's policy is

- To encourage local authorities to bring forward local congestion charging schemes for their areas if they wish to, allowing them to retain the revenue for expenditure on city transport and environmental schemes.
- To consider the case for new sections of tolled inter-urban motorway, following the experience with the M6 toll road around Birmingham.
- To study the technical and economic options for national network road user charging.

On the first of these, the famous example is the London Congestion Charging scheme implemented in February 2003. This is generally considered to have been successful in improving traffic, environmental conditions and bus service quality in Central London, but the scheme has high operating costs. So far, no other city has followed London's example, and a referendum in Edinburgh found against the proposal there by a margin of 3:1. While the Government's Ten Year Plan of 2001 envisaged local road user charging schemes in several cities by 2011, that now seems most unlikely.

On the second point, the government is currently consulting on a new toll motorway between Birmingham and Manchester, and it is believed that other such schemes are in the pipeline. The new interest in toll roads is not only about finance, it is also about providing and maintaining quality of service, 'locking in the benefits' by allowing toll levels to help manage congestion. Some commentators see a wider policy interest in getting the British public accustomed to 'user pays' as a precursor to network road user charging (Mackie & Marsden, 2004).

It is widely believed that neither local charging schemes nor motorway tolls are really the answer to the problems of traffic and network management in the congested parts of Britain. If this problem is to be addressed through pricing, the approach needs to be a network-wide one, which is not restricted by local authority boundaries or limited to specific types of road. This is a very challenging agenda for the government and will certainly require a rebalancing of the existing tax tariff by introducing a new congestion-related element and revising existing charges. A comprehensive report was commissioned by Government in 2004 (Department of Transport, 2004; see also House of Commons Transport Committee, 2005), which concluded that the critical path to delivery of network charging would require at least 10 years from a decision in principle to proceed. It will therefore be very interesting to see how fast the new government moves forward with this agenda which, as other countries such as the Netherlands have found, is economically desirable but politically difficult to deliver.

NOTES

1. This section draws heavily on Bain and Wilkins (2002).

REFERENCES

Bain, R., & Wilkins, M. (2002). *The evolution of DBFO payment mechanisms: One more for the road*. London: Standard and Poors Infrastructure Finance Note, 15/11/202.

Charlesworth, G. (1984). *A history of British motorways*. London: Thomas Telford Ltd.

Currie, D. (2000). *Funding the London underground*. Working Paper no. 35, London Business School, London.

Department of Transport. (2004). *Feasibility study of road pricing in the UK*. 20 July 2004, available at www.dft.gov.uk

Edwards, P., Shaoul, J., Stafford, A., & Arblaster, L. (2004). *Evaluating the operation of PFI in roads and hospitals* (238 pp.). Research Report, Association of Chartered Certified Accountant (ACCA) 84.

HC 844. (1998). *The departmental annual report*. House of Commons Select Committee on Transport, HMSO, July.

Heald, D. (1997). Privately financed capital in public services. *Manchester School, 65*(5), 568–598.

Heald, D. (2003). Value for money and accounting treatment in PFI schemes. *Accounting, Auditing and Accountability Journal, 16*(3), 342–371. Available also at http://www.shef.ac.uk/management/dah/pfivfm.pdf

House of Commons Transport Committee. (2005). *Road pricing: The next steps. Seventh Report of Session 2004/5. HC 218*. London: Stationery Office.

Mackie, P., & Marsden, G. (2004). *M6 toll consultation response*. Mimeo.
Ministry of Transport. (1968). *Road track costs*. London: Ministry of transport.
National Audit Office. (1998). *The private finance initiative. The first four design, build, finance and operate roads contracts.* HC476. London: Stationery Office.
Sansom, T., Nash, C., Mackie, P., Shires, J., & Watkiss, P. (2002). *Surface transport costs and charges*. ITS, University of Leeds, in association with AEA Technology Environment.
Standing Advisory Committee on Trunk Road Assessment. (1994). *Trunk roads and the generation of traffic*. London: Stationery Office.

THE PRIVATE FINANCE INITIATIVE: THE UK EXPERIENCE

Malcolm Sawyer

INTRODUCTION

The essential features of the private finance initiative (PFI) in the UK are that capital investment projects (for the public sector) are financed as well as constructed by a private company, and then leased back to the public sector over a pre-determined period (generally 25–30 years), and that generally the private company provides a range of services associated with the capital project (e.g. maintenance). The PFI involves not only drawing upon alternative sources of finance for public investment but also that services related to the capital thereby constructed are also provided under the PFI contract. The idea of comparing a PFI project with a 'conventional alternative' (labelled public sector comparator, PSC) financed through bonds with services provided by the public sector is central to the operation of PFI.

Much of the impetus for the PFI arose from a combination of a perception of low and declining levels of public investment, concerns over the size of the budget deficit and to some degree of the public debt and allied to this perceptions that government would be unable to borrow further. The arguments for the PFI have evolved over time, and have tended to move away from a stress on the provision of additional finance and to use arguments

Procurement and Financing of Motorways in Europe
Research in Transportation Economics, Volume 15, 231–245

ISSN: 0739-8859/doi:10.1016/S0739-8859(05)15018-5

based on who is best able to bear risk, the relative efficiency of PFI provision with 'conventional' public investment and the incentives involved. This is represented by the following recent statement. "The involvement of private finance in taking on performance risk is crucial to the benefits offered by PFI, incentivising projects to be completed on time and on budget, and to take into account the whole of life costs of an asset in design and construction. Private finance in PFI, particularly third-party finance, takes the risks in a project and allocates them to the party best able to manage them. The lenders to a PFI project, as they have significant capital at risk, have a powerful incentive to identify, allocate and ensure the effective management of all the risks the private sector assumes in a project" (H M Treasury, 2003a, p. 10).

PFI has generally been concentrated in the area of transport, defence, health (often hospital construction) and education (school construction and refurbishment). In the current financial year (2005/06), the estimated capital spending by the private sector under the PFI is projected to be £3,229 million, with projects at the preferred bidder stage amounting to £5,280 million (figures taken from H M Treasury, Budget Report 2005, Table C17). In relation to overall public expenditure the figures appear relatively small: overall public expenditure is estimated at just around £503 billion for 2005/06, and hence PFIs were equivalent to under 1% of public expenditure. A more appropriate comparator would be other forms of public investment: gross investment in 2005/06 is projected to be £47 billion, and after depreciation and asset sales, net investment at £26 billion.

Under the PFI, the government is contractually committed to lease the project from the private sector company for a specified period (often 25–30 years) ahead (and on the other hand, the private company is contractually obliged to lease the project to the public sector). For the private company this provides a guaranteed future income stream (usually in real terms). For the government, there is the contractual obligation to make those future payments. The future obligations amount on an annual basis to around £7 billion (in 2005 prices) for the next decade or more. The cumulative figures give a future commitment of £138.4 billion. A (real) discount rate of 3.5% would put the present value of those commitments at £99 billion and a 6% rate £81 billion. The 3.5% rate was chosen as the newly recommended rate for project appraisal (H M Treasury, 2003b, Annex 6) and 6% as a figure often used in PFI evaluations. By way of comparison, the public sector net debt stood at £415 billion in March 2005 (and £482 billion calculated on the Maastricht treaty definition).

ACCOUNTING ISSUES AND THE SIZE OF PUBLIC DEBT

Two accounting-related issues arise which serve to obscure the role of the PFI. The first issue arises from the concern over the structure of the balance sheet of government which focuses on liabilities and not on assets. In the public sector, assets do not usually directly generate income for the government, which may explain a focus on liabilities rather than assets. But it does ignore that public sector assets are productive in a more general sense and add to the productive potential of an economy and thereby to future tax revenue. Assets should be included in any evaluation of the government's balance sheet position, rather than merely to focus on the liability side.

The second is that the liabilities of government only include the liabilities incurred in the form of financial assets and do not include commitments to future payments. Hence, it neither includes commitments to future transfer payments, such as pensions, nor (and more significantly here) does it include commitments to future payments under leasing agreements. Private sector accounting practice would include leases over 3 years as a future liability in the capital accounts (balanced by a corresponding asset), public sector accounting does not.

The PFI appears to place an investment project 'off balance sheet'. As PFI involves the creation of both assets and liabilities, the net position is unaffected, but having projects which can be placed off balance sheet may influence behaviour. Two sets of accusation have been made in this regard. First, that there has been a drive to get projects 'off balance sheet' in order to limit the apparent size of the government's budget deficit. Second, that some projects may 'disappear', appearing in neither the government's nor the private company's balance sheet.

The UK government has recently denied that these balance sheet considerations have played any role in the development of the PFI. "The decision to use PFI is taken on value for money grounds alone, and whether it is on or off balance sheet is not relevant. Almost 60 per cent of PFI projects by value are reported on Departmental balance sheets and fully reflected in the Government's national accounts. The Government publishes a complete statement of the costs of PFI facilities, which are fully covered by annual unitary payments, in the Budget document." (H M Treasury, 2003a, p. 13). Similarly, "the Government only uses PFI where it offers value for money, considered over the long term. ... The financial reporting and balance sheet treatment of projects are subsequent and irrelevant to the decision whether to use PFI, but the monitoring and reporting of financial commitments made under PFI is an important part of managing the public finances" (H M Treasury, 2003a, b, p. 22).

We would argue that the manner in which PFI is treated in government accounts has generated incentives to use PFI in an era where reducing budget deficits has been high up the political agenda. However, the future impact of PFI may well be to reduce rather than increase public sector expenditure on investment as the expenditure under PFI mount up.

ADDITIONAL INVESTMENT?

It is often argued that the PFI provides additional investment for the public sector. For example, Paul Boateng, then Financial Secretary to the Treasury, argued that "for the future we plan 100 new hospital schemes – including 26 new PFI hospitals to be up and running by mid-2005 – eight already up and running, 15 more at various stages of construction. An investment in our nation's health that would not have been possible without finance from the private sector" (Boateng, 2001). The argument that the PFI provides additional finance for the public sector and enables additional projects to proceed is essentially spurious. However public sector investment is undertaken there are resource costs involved in the construction and maintenance of the investment project, and the investment has to be funded whether directly or indirectly. As Sussex (2001) argues, "if we observe that more NHS investment is made now with the PFI than was made before it was introduced, then this is the result simply of a political decision to increase investment. Given the government's current tests of fiscal prudence, there appear to be no current macroeconomic reasons for preferring PFI to Exchequer financing, or for regarding one approach as any more affordable than the other."

The limitation on public sector investment may be thought to be one of finance and funding. An investment project may be financed from general taxation, which would mean that tax revenue is higher (than it would be otherwise) and finance available for private expenditure thereby reduced. When a government decides that capital expenditure is not to be financed through general taxation but through borrowing it faces, in effect, a choice. The public sector can borrow from the private sector and uses the funds obtained to pay for the investment project, or the private sector finances the project directly as through the PFI schemes. In either case, the public sector is borrowing from the private sector. The public sector repays the full cost of the private sector companies which have constructed the investment projects in annual payments over periods of 20–30 years. But it does not provide access to any higher level of funding than would otherwise be the case with public funding. In either case, the public sector faces future obligations – either in the form of future interest payments on the borrowing or in the form of leasing and other charges to the private sector.

RISK TRANSFER

The general idea that any investment carries risks and the notion that appropriate allowance for risk and risk bearing should be made have become central to the debates over PFI. In effect, it is argued that under 'conventional' public sector investment, the government can borrow at a relatively low rate of interest, which is perceived to be risk free, but the government bears the risks associated with the operation of the public sector investment. Under the PFI, the company concerned borrows at a higher rate of interest, which is reflected in the price it charges the government, but the company bears the risks associated with the operation and maintenance of the investment project.

The general approach of the government to the issue of risk transfer is well summarised as: "The appropriate sharing of risks is the key to ensuring value for money benefits in PFI projects are realised. The benefits described above all flow from ensuring that the many different types of risks inherent in a major investment programme, for example, construction risk or the risk associated with the design of the building and its appropriateness for providing the required service, are borne by the party who is best placed to manage them."

The general principles behind the government's approach to risk-sharing in PFI are:

- The government underwrites the continuity of public services, and the availability of the assets essential to their delivery.
- That the private sector contractor is responsible, and at risk, for its ability to meet the service requirements it has signed up to. Where it proves unable to do so, there are a number of safeguards for the public sector and the smooth delivery of public services in place, but the contractor is at risk to the full value of the debt and equity in the project.
- The full value of that debt incurred by the project, and the equity provided by contractors and third parties, is the cap on the risk assumed by the private sector.

Successful PFI projects should therefore achieve an optimal apportionment of risk between the public and private sectors. This will not mean that all types of risks should be transferred to the private sector. Indeed, there are certain risks that are best managed by the government; to seek to transfer these risks would not offer value for money for the public sector" (H M Treasury, 2003a, p. 35).

The issues arising include the following:

The Pricing of Risk

The significance of this is illustrated by the following taken from a report of the Public Accounts Committee of the House of Commons (Table 1).

The comparator included £151 million for additional risk as a measure of the average cost overrun of 24% in public sector managed projects. This figure was the percentage given by the Treasury, which said that it was at the bottom of the range and it arose from a study carried out by a firm of consulting engineers. This risk allowance in itself more than accounted for the difference between the PFI bid and the comparator and had done so in other projects as well.

Other PFI deals had used different, lower risk addition percentages; and with a range of other adjustments available from the use of different discount rates (and the way service costs were spread), it seemed to us that the public sector comparator figures could be used to demonstrate any result required. Such uncertain figures risked clouding the issue of value for money and could cloak a predisposition to go in for PFI (Committee of Public Accounts, 2004).

Others have similarly commented that "in all schemes [considered] risk transfer is the critical element in proving the value for money case. There is considerable variation between schemes in the absolute and relative value of risk transferred. What is striking, however, is that in all cases risk transfer almost equals the amount required to bridge the gap between the public

Table 1. The Final Comparison between the PFI Bid and the Public Sector Comparator Net Present Value (£ Million).

	IAS Deal	Public Sector Comparator
Building, refurbishment and services	489	605
Risk adjustment		151
Technical transition	264	68
Total	753	824

sector comparator and the PFI. This suggests that the function of risk transfer is to disguise the true costs of PFI and to close the difference between private finance and the much lower costs of conventional public procurement and private finance" (Pollock, Shaoul, & Vickers, 2002, p. 1208).

It can be argued that "risk transfer requires the ability to quantify the probability of things going wrong. There is no standard method for identifying and measuring the values of risk, and the government has not published the methods it uses. The business cases we examined do not reveal how the risks were identified and costed. Our findings are supported by a Treasury commissioned report which found that in over two thirds of the business cases for hospital PFI schemes the risk could not be identified. In the other cases risk transfer was largely attributed to construction cost risks, which would be dealt with by penalty clauses under traditional procurement contracts" (Pollock et al., 2002, p. 1208).

It is 'Risk' Effectively Transferred from Public Sector to Private Sector

One way in which risk is transferred from the public sector to the private sector under PFI is that the payments under the contract are assured and the contractor accepts the risks associated with the provision of the services under the contract which include variations in costs and impact on profits. The effective transfer of risk would, of course, mean that in the event of a major difficulty which threatened the profitability of the PFI project, there would be no assistance forthcoming from the government but the PFI contractor would have to bear the costs. There is, not surprisingly, problems arising here from the 'too big to fail' syndrome. An example of this is given in the following: "The Passport Agency PFI provides an example of the political realities of risk transfer in the context of a high profile, essential service. The fact that compensation was waived and the allocation of the costs of failure negotiable suggest that risk transfer was not after all secured by the contract, or not to the value contractually specified and in respect of which the risk premium was payment" (UNISON, 2004, p. 33).

"In conclusion, our analysis has shown that the concept of risk transfer that lies at the heart of the rationale for partnerships is problematic, regardless of whether the project is 'successful' or not. If the project is successful, then the public agency may pay more than under conventional procurement: if it is unsuccessful then the risks and costs are dispersed in unexpected ways. Hence public accountability is obscured. ... our analysis shows that, although a project fails to transfer risk and deliver value for money in the way that the public agency anticipated, the possibility of enforcing the arrangements and/or dissolving the partnership is in practice severely circumscribed for both legal and operational reasons" (Edwards & Shaoul, 2002, p. 418).

"[R]isks can be transferred only through a contract that identifies them. Yet there is reason to cast doubt on the claim that contracts offer a means of transferring financial risk. Where a trust wishes to terminate a contract, either because of poor performance or insolvency of the private consortium, it still has to pay the consortium's financing costs, even though the latter is in default. It would otherwise have to take over the consortium's debts and liabilities, given that the lending institutions make their loans to the consortiums conditional on NHS guarantees. In such cases 'the attempt to shift financial responsibility from the public to the private sector fails. De facto, a risk-sharing arrangement results from force majeure', as the Railtrack collapse has shown" (Pollock et al., 2002, pp. 1208–1209).

Is There a Net Change in the Amount of Risk?

The first point here is that when risk is considered in a probabilistic manner and when there is no significant (differential) impact of risk on behaviour, then the public sector benefits from 'the law of large numbers' in terms of risk. Having the projects undertaken on an individual basis and with the risk of individual projects separately priced, then the degree of risk is then greater for the separate schemes than for the pooled arrangements.

The second is that for the public sector there is a potential loss of flexibility under the PFI arrangements. Any contract drawn up for any significant period of time suffers from issues of flexibility in so far as the contract cannot possibly specify reactions to all changes in circumstance. Seeking to do so would entail extremely long contracts, and may not be possible in a world with some degree of uncertainty. The PFI contracts are typically for 25–30 years, and specify the services to be rendered over that period. The 'conventional' public sector alternative also includes degrees of inflexibility: once a school is built, its use cannot be readily changed, etc. But it is clear that there is some flexibility: demographic changes may render the school surplus to requirements, its use may be changed and the associated maintenance arrangements, etc., changed. Under a PFI, compensation to the contractor would be required, etc. "PFI, by locking management into a particular form of service delivery and one contractor for 30 years, serves to reduce rather than enhance management's flexibility to respond to changed circumstances, as other analysis has shown" (Shaoul, 2005, p. 450).

Thus, it can be readily argued that the PFI arrangements reduce flexibility and inhibit responses to changing circumstances. In that regard PFI arrangements increase overall risk rather than diminish it.

COST OF FINANCE

A PFI scheme in effect replaces direct borrowing by government with indirect borrowing, and moves from government being able to borrow at the lowest rates (on government bonds) to borrowing through more expensive channels. This view that the cost of finance under PFI is higher is to some degree countered by the argument that there is a shifting of risk and the private sector is bearing the risk – at a cost reflected in the higher cost of capital.

The Treasury have sought to address this argument that PFI finance is more expensive than 'conventional' public investment finance when they argue that

> "there are two common assertions made to justify the claim that the private sector cost of capital exceeds the public sector cost of capital:
>
> 'Governments can borrow at a risk free rate of interest'
>
> This is not the case, there is a risk premium either way, it is just explicit in the price of private capital. Where gilts are used, taxpayers effectively underwrite the associated risk and the price reflects this fact. The taxpayer takes on the contingent liability, and where the risk materialises, they carry the cost as a result. If the taxpayer were to be compensated it would be equivalent to paying the risk premium at the point of raising the capital, making the public and private sector's cost of capital equivalent.
>
> 'The government are better at diversifying the risk than the private sector'
>
> This assertion is based on the Arrow–Lind theorem, an academic theory which assumes that project returns can be treated as wholly independent of National income. In fact this is rarely the case as public investment is not risk free" (H M Treasury, 2003a, p. 124).

The Treasury estimate that "since 1995, this 'all-in' cost of private finance has fallen from 13.5 per cent to just under 10 per cent by 2001" (H M Treasury, 2003a, p. 123), which would be significantly above the interest rate on bonds. Although the cost of private finance fell, it is also the case that the interest rate on government bonds also fell over this period by around 3 percentage points.

The Treasury then argues that

> "The study [by PricewaterhouseCoopers] suggests that the most appropriate benchmark to use for the WACC [Weighted Average Cost of Capital] is the regulated utility sector. Further assumptions used in the study are the risk free rate of return, being the long term

government bond yield, and the equity market risk premium, which is assumed to be 5 per cent" (H M Treasury, 2003a, p. 126):

- The average spread between the project IRR [internal rate of return] and benchmark WACCs has been some 2.4 per cent in total;
- On our assumptions about unsuccessful bid costs this reduces to between 1.1 per cent and 1.7 per cent – say 1.4 per cent. To the extent that the assumptions on bid costs are changed it affects conclusions on the allocation of the spread but not the total figure of 2.4 per cent;
- Some 0.7 per cent of the spread is explained by swap costs;
- After considering other factors, we believe the other 0.7 per cent indicates excess projected returns to investors, and that this is due to structure issues that limit competition in the PFI market;
- Bidders' target equity returns average 14.5 per cent over the period before adjustment for bid costs, whereas the cost of equity implied by a traditional WACC calculation is in the range 8.3 per cent–9.4 per cent depending on the assumptions used;
- There is some evidence that spreads were increasing until 1998 but that since then this has reversed;
- Changes in the general capital market environment – such as declining interest rates and margins – have been reflected in PFI financings to the benefit of the public sector;
- We expect the trend towards reduced returns to continue. The effects of steps already taken to standardise processes and share market information have not yet been fully reflected in closed deals because of the length of the procurement process" (H M Treasury, 2003a, p. 127).

Hence, the cost of capital is around 8–9% and the internal rate of return is significantly above the cost of capital; on that basis the cost of capital is substantially above the bond rate, and then there is a further 'premium' being paid to the PFI contractors.

The PwC study identifies certain features of PFI which could give rise to such a spread:

- the risks of political intervention given the novel nature of this form of contract, the long term nature of the commitment and perception of political risk;
- corporate investors may use corporate hurdle rates, based on their core business on pricing their investors, which in PwC's view will nearly always be higher than is appropriate for PFI projects;
- returns may be influenced by the requirements of debt funders and their cover ratio requirements;
- long periods of negotiation which result in financing terms being agreed early, but closed at a later date after financial markets have moved favourably ([5]H M Treasury, 2003a, p. 127).

Another aspect which points towards the additional costs of finance associated with PFI comes from the re-financing of PFI projects. The risks associated with a proposed PFI project change between the pre-contract period (when there is still uncertainty over the award of the contract as well as its precise terms) and the post-contract period. This can then be reflected in a lower cost of finance (for the PFI contractor) in the post-contract period than in the pre-contract period. The ability to re-finance the PFI contract can be an additional source of profit for the contractor. "[O]nce the required service has been brought into operation, the project risks are lower, as the risks associated with commencing service delivery are no longer relevant. This creates opportunities to reduce the annual financing costs, as funders are prepared to offer better terms for projects with lower risks. ... Lower annual financing costs improve the returns that can be paid to the private sector shareholders" (National Audit Office, 2002, p. 1). But "Since June 2001, most new PFI deals have included arrangements to share refinancing benefits 50/50" (National Audit Office, 2002, p. 23).

It would then appear that the cost of finance is indeed higher under PFI than for 'conventional' public investment, though the extent of the difference and how far the additional cost is justified by risk transfer may be a matters of debate.

EFFICIENCY GAINS?

As part of the process leading to the adoption of a PFI project, a comparison has to be made between the proposed PFI and a PSC. The claims that the PFI projects are more cost effective than 'conventional' public sector projects have been examined by reference to the comparisons between PFI and PSC. In making use of these comparisons the obvious problem arises: the PFI schemes which proceed are those which are believed to be more cost effective, whereas there are some potential PFI schemes which do not proceed as they are judged to be less cost effective.

> "The Government's policy is to use PFI only where it represents the best procurement option and as shown above, this is unlikely to be the case for projects with a small capital value. It is important then that local authorities have the flexibility to develop such projects through a wide range of procurement routes, choosing the most appropriate option that delivers the best value for the project. This flexibility is part of a wider commitment to devolve responsibility to local councils to meet local priorities, increase local choice and improve performance by removing unnecessary controls that stifle local innovation."
>
> (H M Treasury, 2003a, p. 87)

Given the pressures to adopt PFI rather than 'conventional methods', there are clear incentives to overstate the costs of the PSC alternatives. Further, the PFI is carried through whereas the PSC is not, making a genuine comparison fraught with difficulties.

> "There is a demonstrated, systematic, tendency for project appraisers to be overly optimistic. This is a worldwide phenomenon that affects both the private and public sectors (Flyvbjerg, Underestimating Costs in Public Works Projects – Error or Lie, *APA Journal*, 2002). Many project parameters are affected by optimism—appraisers tend to overstate benefits, and understate timings and costs, both capital and operational.
>
> To redress this tendency, appraisers should make explicit adjustments for this bias. These will take the form of increasing estimates of the costs and decreasing, and delaying the receipt of, estimated benefits. Sensitivity analysis should be used to test assumptions about operating costs and expected benefits."
>
> (H M Treasury, 2003b, paras 5.61 and 5.62)

Two particular issues have arisen while comparing PFIs and PSCs. The first concerns the appropriate discount rate to use. The time profiles of the costs involved are rather different – the PSC is 'front loaded' with the capital costs incurred in the initial stages, whereas for the PFI as far as the government is concerned the costs are more evenly spread out as the leasing charges repay the capital costs of the constructor. In view of these marked different time profiles, the comparison between PSC and PFI may be rather sensitive to the choice of discount rate. The difference in NPV terms between a PFI and the corresponding PSC may be rather small, and hence a relatively small change in the discount rate could well lead to a change in the relative ranking. Further, it may be noted here that the government has recently lowered the test discount rate from 6 to 3.5%, and many PFIs had used the higher 6% rate in the calculations.

The second issue arises from the treatment of risk. It has been seen above that the notion of the transfer of risk is an important feature of PFI. Since it is argued that under the 'conventional' public sector project much risk is borne by the public sector whereas under PFI the bearing of risk is transferred to the private sector, then allowance for the costs of bearing risk should be made when the PSC is drawn up. But how that risk is assessed and measured becomes significant in the judgement between the PSC and the PFI, and yet the measurement of risk is problematic.

The strongest argument for the cost effectiveness of PFI has come from studies such as Arthur Andersen and Enterprise LSE. They report, for example, that

> "The table below lists the high level value for money data we extracted from 29 FBCs received following a request from TTF to departments.

Four figures are quoted:

1. The Net Present Cost ('NPC') of the PSC. This is the net cost (taking into account any project revenues) estimated by the public sector of undertaking a project itself and producing the same or similar outputs under conventional procurement. The NPC should include an estimate of the risk that would be retained by the public sector compared to the PFI option.
2. The NPC of the PFI option. This is the cost to the public sector of making payments to the service provider over the life of the contract. The payment profile should assume that no deductions are made for poor performance.
3. The estimated saving to the public sector in NPC terms of entering into the PFI contract. This is the difference between the NPC of the Public Sector Comparator and the NPC of the PFI option.
4. The estimated cost saving as a percentage of the NPC of the Public Sector Comparator.

The total estimated saving from our sample of projects is over £1 billion in NPC terms against an estimated cost of conventional procurement of £6.1 billion. The table shows a consistent pattern of PFI projects delivering sizeable estimated cost savings. The average percentage saving for this sample of 29 projects (i.e. the percentages above added and divided by the number of projects, a calculation that avoids the large projects distorting the average) is 17%. This compares to the average saving in PFI projects with a PSC examined to date by the NAO of 20%. On the basis of the public sector's own figures, the data therefore suggests that the PFI offers excellent value for money.

Of course, the headline PSC savings form only part of the picture. As we noted above, the jury is still out on the extent to which PFI contracts will deliver the benefits promised. Where projects are not 100% successful then the apparent benefit of the value for money saving will be diminished. This point has been demonstrated most visibly in the IT sector where there have been a number of high profile problems in delivering against the original specification"(Arthur Andersen and Enterprise LSE, 2000).

However, it has also been noted that "The estimated basic construction costs in the final Comparator were increased by 24 per cent in line with Treasury advice on historical cost overruns on large scale public sector projects. ... As in other PFI cases, the adjustment for risk on construction costs of the public sector alternative more than accounts for the estimated cost difference between the comparator and the PFI deal" (National Audit Office, 2003, p. 25).

The basis of this argument is that there have been cost overruns in past 'conventional' public investment projects but that similar overruns either do not occur under PFI or if they do the consequences are absorbed by the contractors rather than the government. It can be first noted that for some of the examples given above cost overruns do occur under PFI arrangements. The recent withdrawal (voluntary or otherwise) of Jarvis from a range of PFI contracts in the face of substantial financial losses may suggest that there are cost overruns. As the method of calculation of the 24% figure cited above has not been revealed, we cannot be sure that a like-for-like comparison is being undertaken: the costs of public investment can be affected by inflation, changes of specification, etc. But the main point here is that the main argument for lower costs under PFI arises from the nature of the contract. The PFI contract is in effect a fixed price contract which can also contain bonuses for early completion and cost penalties for late completion. These features could readily be incorporated into 'conventional' public investment contracts, which would reduce (or remove) any cost advantages of the PFI arrangements.

CONCLUSIONS

This chapter has outlined some of the issues involved in the evaluation of the PFI experience in the UK. We have argued that if PFI does give rise to additional investment, then that comes not from finding some additional sources of finance but rather through a political decision which views equivalent expenditures on PFI and on 'conventional' public sector investment from a different accounting perspective. Further, the cost of finance under PFI is likely to be greater than it is under 'conventional' public investment. This means that the cost in terms of public expenditure of PFI is that significantly greater than 'conventional' public investment.

The claimed efficiency gains of PFI over 'conventional' public investment appear to arise predominantly from the pricing of risk in the PSC and from the perceived overrun of costs under 'conventional' public investment.

REFERENCES

Arthur Andersen and Enterprise LSE. (2000). *Value for money drivers in the private finance initiative* (from web site http://www.ogc.gov.uk).

Boateng, P. (2001). Keynote address. In: *Public private partnerships/private finance initiative global summit*, Dublin, 16 October.

Committee of Public Accounts. (2004). *Government Communications Headquarters (GCHQ): New accommodation programme twenty–third report of session* 2003–04. HC 65.

Edwards, P., & Shaoul, J. (2002). Partnerships: for better, for worse? *Accounting, Auditing & Accountability Journal, 16*(3), 397–421.

H M Treasury. (2003a). *PFI: meeting the investment challenge*. Accessed on http://www.hm-treasury.gov.uk

H M Treasury. (2003b). *The green book appraisal and evaluation in central government*.

National Audit Office. (2002). *PFI refinancing update*. HC 1288 Session 2001–2002.

National Audit Office. (2003). *Government Communications Headquarters* (GCHQ): *New accommodation programme*. HC 955, July.

Pollock, A. M., Shaoul, J., & Vickers, N. (2002). Private finance and 'value for money'. In NHS hospitals: a policy in search of a rationale? *British Medical Journal, 324*, 1205–1209.

Shaoul, J. (2005). *A critical financial analysis of the private finance initiative: selecting a financing method or allocating economic wealth? Critical Perspectives on Accounting, 16*(4), 441–471.

Sussex, J. (2001). *The economics of the private finance initiative in the NHS*. Office of Health Economics: Summary on http://www.ohe.org (accessed 9 May 2002).

UNISON. (2004). *Public risk for private gain? The public audit implications of risk transfer and private finance*. Report research written for UNISON by Allyson Pollock, David Price, and Stewart Player.